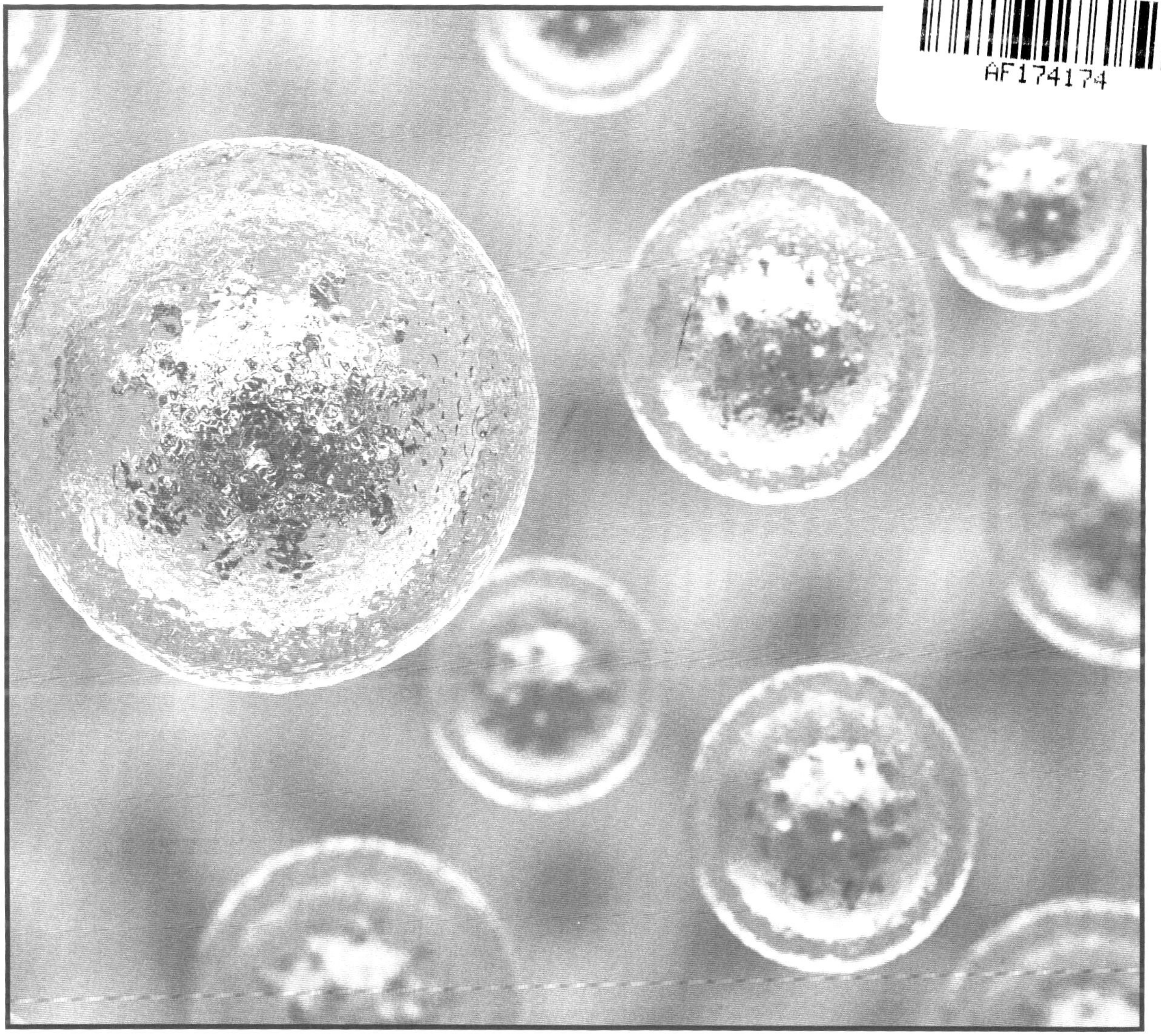

BIO 1120

A Laboratory Perspective

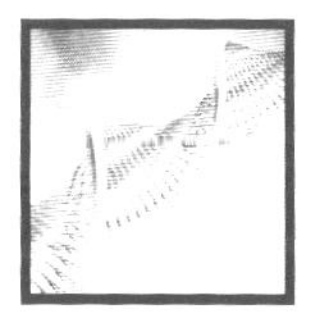

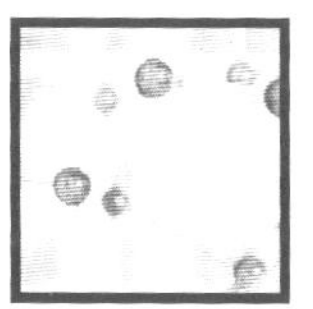

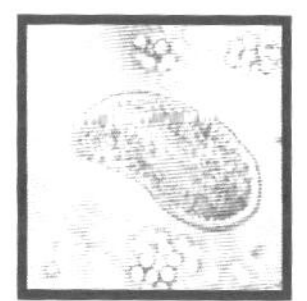

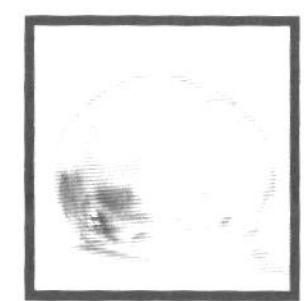

Wright State University
Biology Department
Dan Krane | 2019–2020

VAN-GRINER
LEARNING

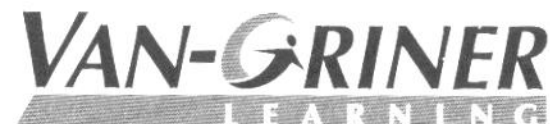

BIO 1120
A Laboratory Perspective

Wright State University
Biology Department
Dan Krane | 2019–2020

Printed in the United States of America
10 9 8 7 6 5 4 3 2 1
ISBN: 978-1-61740-745-1

Van-Griner Learning
Cincinnati, Ohio
www.van-griner.com

President: Dreis Van Landuyt
Project Manager: Janelle Lange
Customer Care Lead: Lauren Houseworth

Krane 745-1 Su19
312610
Copyright © 2020

TABLE OF CONTENTS AND SYLLABUS

SYLLABUS

The laboratory experiments you will be performing are an integral part of BIO 1120, and each has been carefully designed to complement the material presented in the lecture portion of the course. If an emergency prevents you from attending your regular laboratory section, *you must make up the work* in another laboratory section *during the same week.* Be certain to ask the graduate teaching assistant for the lab section you would like to attend for permission to participate in their section's activities that week.

Evaluation of your work in the laboratory portion of BIO 1120 will account for 25% of your overall grade for the course **(the percent of 130 possible points for the lab will be what is carried over for the overall grade for the course).** Laboratory reports must be turned in for Experiments 3, 6, 7, 8, 11, and 13 and will be due at the beginning of the next week's experiment. Each report must follow the guidelines and length restrictions described on the back of this page and will be worth 10 points. Late reports will be docked at the rate of one point per weekday that they are late. Answers to questions at the end of Experiments 1, 2, 4, 5, 9, 10, and 12 must also be turned in at the start of the next week's experiment and will also be worth 10 points each. All laboratory reports and answers to questions will count toward your grade (none will be dropped).

The two experiments marked with asterisks (*) in the Table of Contents will have you using expensive microscopes. Please make sure to read the instructions in the lab manual on the safe use and handling of these microscopes.

Each section of the BIO 1120 laboratory will have a designated graduate teaching assistant. The teaching assistant for your section will announce a time and place for their office hours. Your teaching assistant and their office hours are extremely valuable resources for both the laboratory *and* lecture portions of BIO 1120.

LABORATORY REPORT GUIDELINES

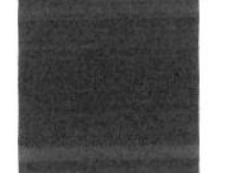    

GUIDELINES

Scientists are constantly called upon to communicate their results in written form. Abstracts for talks at scientific meetings, published research reports, and grant applications are just a few of the many forums in which scientists must write effective descriptions of their studies and conclusions.

Written scientific material usually has a distinguishing style and is typically divided into five separate components. Each of the laboratory reports you prepare for BIO 1120 should have the parts listed below and will be awarded the following points by your section's teaching assistant:

Section	Length	Points
Title	Not more than one sentence	1
Introduction	One or two paragraphs	2
Materials and Methods	Usually one paragraph	2
Results	Three to five paragraphs, tables, or graphs	2
Conclusions	Two to four paragraphs	3
		10

Your laboratory report's **title** should succinctly describe the nature of the experiment that you have performed. "Hydronium Ion Concentrations and Buffers" would be a suitable title for your first report. "Experiment 1" would not.

The **introduction** of a laboratory report should convey the purpose of the experiment and describe in general terms the approach that you employed to address the scientific questions. It should also include a summary of the significance of previous work that others have performed using the same methodology or in the same field.

The **materials and methods** section of your report should have within it all the information that another scientist would need to be able to reproduce your results. In many cases reporting "The experiment was performed as described in section X of the BIO 1120 laboratory manual" will be sufficient. However, be certain to note *any* modifications of the procedures (even those that were not intentional) and the specific materials that you used when the manual gives you the opportunity to make choices.

All the observations you make during the course of your experiment as well as tables and graphs that summarize your data should appear in the **results** sections of your reports. The laboratory manual itself will prompt you to record (and, subsequently, report) most relevant observations and data.

The **conclusions** section of your report is the only portion in which you should discuss the implications of your results and comment on their reliability (not on your personal feelings about the experiment you performed). Any factors that may have resulted in departures from the results that were expected should be discussed and theories that have been supported (or disproved) should be listed in this part of your reports.

A sample laboratory report for the first experiment (Concentration, pH, and Buffers) is included at the end of this laboratory manual for you to use as a guide. All laboratory reports for this course will be submitted electronically and evaluated for originality. You will be able to see if parts of your report look suspiciously similar to reports that other students have previously turned in before you actually turn your report in. If you are relying on other sources for your report, be certain to use quotes and provide proper citations.

PRELIMINARY CONSIDERATIONS

"In science, as in life, learning and knowledge are distinct, and the study of things, and not of books, is the source of the latter." Thomas Henry Huxley (1825–1895)

CONSIDERATIONS

Our understanding of living systems in all their complexity is based on the scientific enterprise of making observations, asking answerable questions, and testing hypotheses. Participating in the process of science and developing creative and critical reasoning abilities is an indispensable aspect of learning biology.

The laboratory exercises for BIO 1120 have been designed to illustrate concepts central to the lecture portion of the course and biology in general. Hopefully, you will find them all interesting and topical in their own right as well! Whenever possible, you should strive to apply the insights you have gained in the laboratory to the material you are studying in the lecture portion of the course.

The fourteen sets of experiments that follow will encourage you to synthesize results from observations and experiments and then draw conclusions from evidence in the truest fashion of the scientific method. Some of this work may make you want to apply your results to new problems or pursue open-ended follow-up investigations of your own. If it does, you will have experienced some of the thrill associated with discovery—share your ideas with your section's teaching assistant or the course's instructor, and we'll do what we can to set you up with the resources you need to pursue your interests.

SAFETY

Accidents of almost any sort are rare in biology laboratories at Wright State University. This is the fortunate consequence of careful design of the experiments and also the careful attention of teaching assistants, laboratory prep personnel, and most students. Nevertheless, the hazard of accidents exists at all times. Attention is called to that possibility here so that you will be aware of how important safety is in *any* laboratory at *any* time. The best general advice regarding safety that can be imparted is twofold:

1. **Read the directions given in every experiment carefully and in advance.** Know precisely what you are going to do before you come into the laboratory. Think about what procedures might be hazardous.

2. **Use common sense.** If something looks dangerous, it probably is. If someone near you appears to not be using good judgment, help them out. In keeping someone else from having an accident you will have performed a very valuable service not only to them but to everyone else in the laboratory as well.

Experiments in biology laboratories sometimes involve potentially dangerous chemicals and expensive equipment. To ensure your personal safety and the integrity of the equipment you use, follow these simple rules while you are in the laboratory:

1. **Do not eat, drink, or smoke in the laboratory.** In "Experiment 10: Human and Population Genetics" you will perform this semester your teaching assistant will give you the opportunity to eat a small amount of asparagus as part of a test of your genetic make-up. Special precautions will be in place at that time but at no other. This ban applies to gum chewing and nail-biting as well.

2. **Protect your eyes by wearing safety glasses and tie back long loose hair.** Wearing appropriate safety glasses in laboratories of any sort is actually required by law in the state of Ohio and for good reason. Contact lenses are not a substitute for safety glasses. If any chemical reaches your eyes in the laboratory, immediately flush them with an ample amount of water.

3. **Report any spills to your teaching assistant immediately.** Most chemicals are only dangerous if they linger. If you spill anything on yourself, wash it off immediately with an ample amount of water. Know the location of the nearest safety shower so that you can move to it or direct others to it immediately if necessary. Some of the chemicals you will be working with interact adversely with skin and clothing—sandals are **NOT** allowed in the laboratory. Shorts and other clothes that leave large areas of skin unprotected are also **NOT** allowed.

4. **Keep open flames away from flammable chemicals or materials and yourself.** Roll up long sleeves when using equipment or open flames. Know the location of the laboratory's fire extinguisher.

5. **Determine the shortest exit route from the laboratory and building.** Be prepared to leave the room quickly in the event of a fire or at the instruction of your teaching assistant.

6. **Report any accidents, no matter how minor, immediately.** No cut, burn, or spill is too minor to bring to the attention of your teaching assistant. The instructional personnel in the laboratory will be much more knowledgeable about what hazards specific incidents pose to your health and may administer first aid or arrange for medical attention.

In any laboratory environment it pays to be prepared and to keep your surroundings as neat and clean as possible. Preparedness and good housekeeping contribute equally to better laboratory performance and enhanced safety.

EXPERIMENT 1:
DILUTIONS AND STANDARD CURVES

"The beginning is the most important part of the work." Plato

OVERVIEW

This experiment will familiarize you with some of the tools that you will be using over the course of the semester in this laboratory and to also give you practice with analyzing and presenting data.

Students are often surprised at how much work in a biology laboratory seems to be done on faith. The reality is that biologists often work with such small quantities of such small things that they cannot be seen, and a number of steps in a process must be completed before it is possible to visualize a result. It often takes a bit of practice to become comfortable with handling and moving quantities of things that you cannot see and hold.

INTRODUCTION

Research in the biological sciences often requires manipulation of small volumes of liquid. This procedure will allow you to become more familiar with the operation of micropipettes, tools used to handle volumes of liquid 1 milliliter or less. Becoming skilled in the use of micropipettes will be critical to the accuracy of your results throughout this class. Care must also be taken in the handling of micropipettes because they are both expensive and easily broken when used incorrectly.

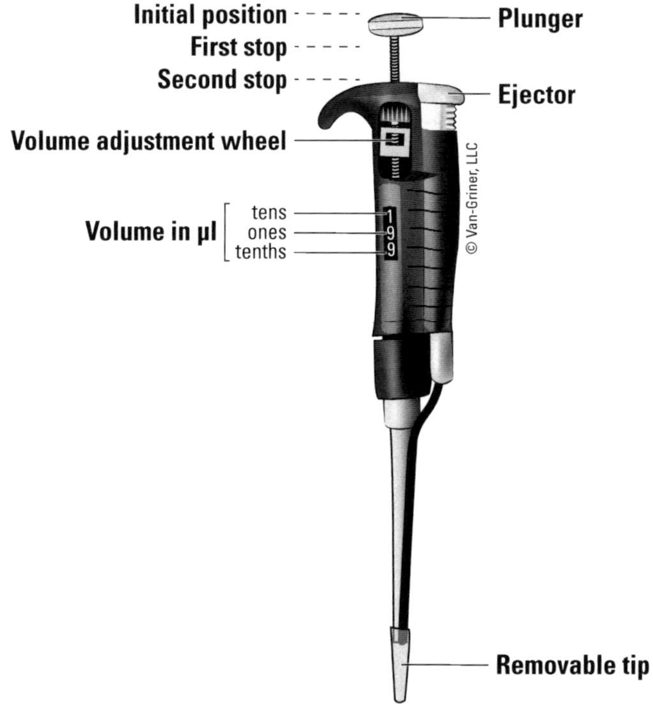

Figure 1.1 Parts of a typical micropipette.

There are several brands and types of micropipettes available, but those that have adjustable volume settings share many key features (Figure 1.1). At the top is a plunger that can be pressed and slowly released to draw up liquid and then pressed again to dispense the liquid. Both of these tasks should be performed using what is known as the "first stop" of the plunger. Occasionally, the second stop of the plunger is used in the dispense stage, but only if liquid remains in the tip of the micropipette.

The "barrel" of a micropipette is the bottom portion of the instrument. The end of the barrel has a location for insertion of a disposal plastic tip. The micropipette may only be immersed in liquid in the area covered by the plastic tip. After use, the plastic tip may be disposed of in a waste container by pushing the tip ejector button, and the ejector arm will press down along the barrel to release the tip.

A dial on the body of each micropipette shows its volume setting. For some micropipettes, the volume is adjusted by turning the plunger, and on others there is a separate volume adjustment wheel. It is very important to know the volume limits of a micropipette before turning the volume adjustment dial. Going above or below the range will cause the instrument to lose calibration and become inaccurate (going too far out of range can even break a micropipette beyond repair). The micropipettes currently used for this lab are shown in Figure 1.2. Note the volume setting limits displayed on the body of the instruments.

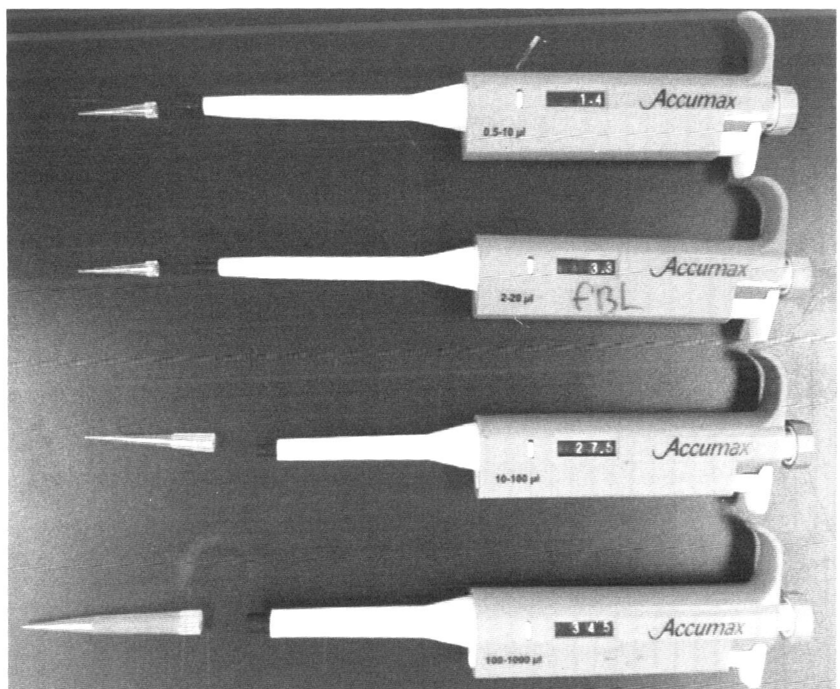

Figure 1.2 From top to bottom, p10, p20, p100, and p1000 micropipettes and corresponding tips.

Once you learn to properly handle the micropipettes, you will use them to make a small dilution series of bovine serum albumin (BSA: a protein widely used in molecular biology laboratories). This means you will use a more concentrated solution of the protein to sequentially make three less concentrated protein solutions. Each dilution will reduce the concentration of the protein by a factor of two. This means that within the series, each dilution is 1/2 of the concentration of the solution preceding it (Figure 1.3). You will then use the series of solutions with known but different protein concentrations in a chromogenic assay where color is produced, the intensity of which depends on the amount of protein present.

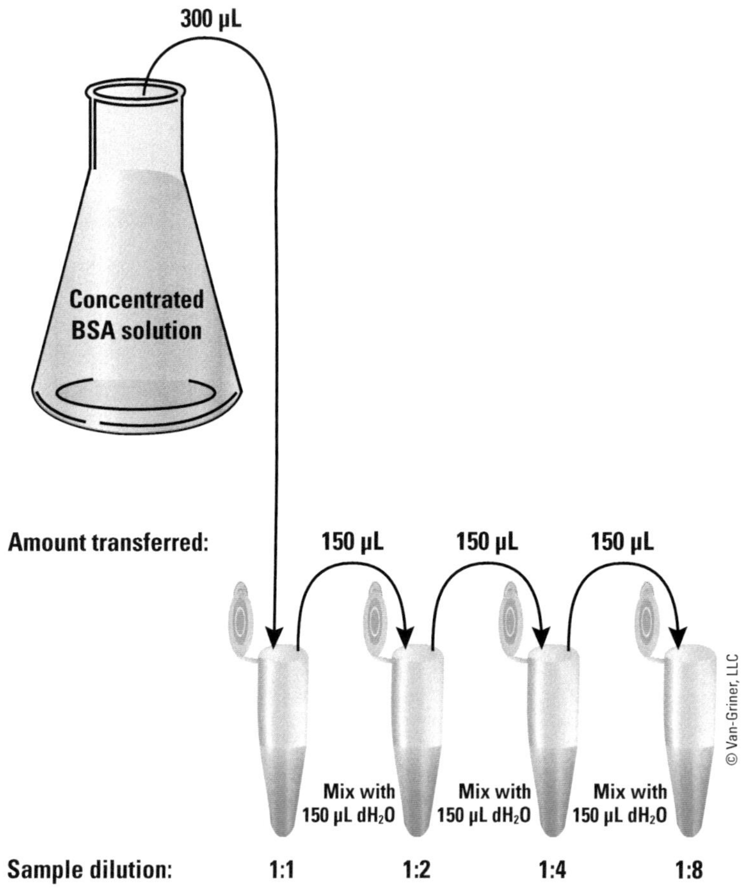

Figure 1.3 Diagram of a serial halving dilution.

The Bradford Assay is a chromogenic reaction that is commonly used to quantify protein amounts. The Coomassie Brilliant Blue G-250 dye that is used in a Bradford Assay can bind to proteins and produce a blue (with a peak absorbance of light at 595 nm wavelengths) product that can be detected by a spectrophotometer. Studies of the dye and various synthetic polypeptides have shown that the dye binds mostly to basic and aromatic amino acid residues of proteins.

In trying to assay samples with unknown amounts of protein, a standard curve (sometimes called a "calibration curve") using known amounts of protein must be generated. It is best if a purified preparation of the protein being assayed is used, but in many cases that is impractical. For such situations and especially if only the relative amount of protein is required for the assay, universal protein standards like BSA (Bovine Serum Albumin) are often used.

The colorimetric assay can be quantified by generating absorbance readings with a spectrophotometer (Figure 1.4). Protein concentration and absorbance should have a linear relationship within a certain concentration range of BSA protein. Therefore, known protein concentration versus absorbance can be plotted to generate a standard curve. The equation for this line can be used to determine the amount of protein in solutions of unknown concentration based on their absorbance readings.

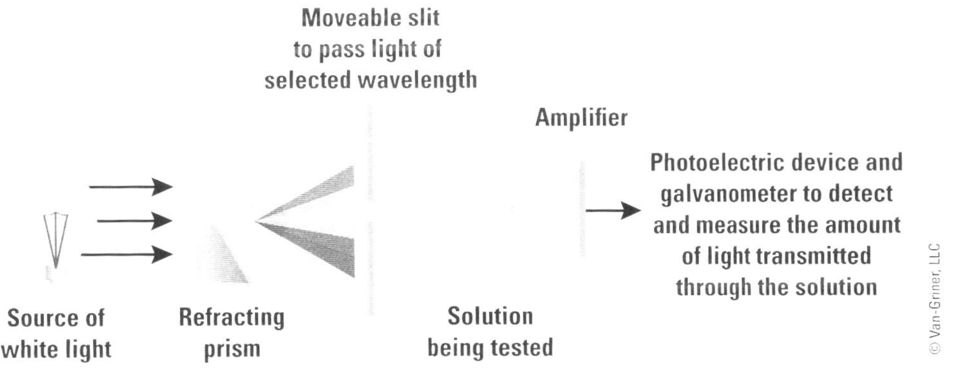

Figure 1.4 Basic components of spectrophotometry.

Practicing the use of micropipettes will aid you in future labs in this course as well as other laboratory-based courses. In particular, you might find it to be helpful to refer back to this experiment when you are generating standard curves as part of "Experiment 6: Enzyme Purification and Assay" for this course.

STEP ONE

Follow along with your GTA to review the parts and functions of the micropipette.

1. Pick up one of the micropipettes on the bench. Push the plunger until you feel the position of the first stop, then slowly release.

2. After engaging and releasing the first stop several times, proceed through the first stop to the second stop of the micropipette, then slowly release.

3. Place a tip of appropriate size on your micropipette, then use the tip ejector to release the tip into the plastic beaker on the bench.

4. Now, adjust the volume of the pipet by turning the dial on top of the plunger. Prior to any adjustment, be sure to read the top and bottom range marked on the micropipette and **DO NOT** go any lower or higher.

5. Set the micropipette to the middle of its range.

6. Place an appropriate tip on the micropipette and practice drawing up distilled water then dispensing it into the plastic beaker by using the first stop only. Repeat until comfortable, using controlled motions. There is no need to change pipet tips between draws at this point.

STEP TWO

Work in groups of four to make a dilution series of the protein BSA and generate a standard curve of concentration versus absorbance following the Bradford Assay.

1. Label three 1.5-mL microcentrifuge tubes as "1:2," "1:4," and "1:8."

2. Set your p1000 micropipette to 150 µL (equivalent to 0.15 mL).

3. Retrieve a tube of the concentrated BSA solution (2 mg/mL).

4. Make a 1:2 dilution of the BSA solution. This means that for every two parts of solution, one of those parts will be concentrated BSA solution and one will be distilled water (dH_2O). To do this, set the p1000 micropipette to 150 µL, place a blue tip on the barrel, and draw up 150 µL of the concentrated BSA. Dispense it into tube "1:2" and add 150 µL of dH_2O. Mix well by pipetting up and down.

5. Next, make a 1:4 dilution of the original concentrated solution. Dispense 150 µL of the 1:2 dilution into tube "1:4" and mix thoroughly with 150 µL of dH_2O.

6. Finally, make a 1:8 dilution of the original concentrated solution. Dispense 150 µL of the 1:4 dilution into tube "1:8" and mix thoroughly with 150 µL of dH_2O.

7. Retrieve five clean cuvettes and label as in point 1 but also include a "C" tube (for "concentrated") and a "B" tube (for "blank").

8. Using a p100 micropipette, place 50 µl of dH_2O into the "B" cuvette.

9. Place 50 µL of each solution from the dilution series into its corresponding cuvette.

10. Add 1500 µL of the Bradford Assay reagent (Coomassie Brilliant Blue G-250 dye) to the blank and to each of the four cuvettes containing 50 µL of BSA solution.

11. Snap the caps onto the cuvettes and invert and flick to mix well.

12. Let the cuvettes sit at room temperature for five minutes.

13. Make sure the spectrophotometer is set to 595 nm absorbance.

14. Wipe the sides of the "B" tube with a kimwipe and use the tube to set the blank for the assay in the spectrophotometer.

15. Wipe down the sides of the remaining cuvettes with a kimwipe and record their absorbance at 595 nm in Table 1.1 below.

16. If time permits perform points 1 through 15 for a second time so as to have a replicate of each data point in Table 1.1.

17. Use the absorbance data and the known concentrations of BSA in each reaction to make a plot of absorbance at 595 nm (Y, vertical, axis) versus protein concentration (mg/mL, X, horizontal, axis). The plot should have a title and labeled axes (with units included).

18. Use Excel or a similar method to generate a line of best fit on the graph. Choose the option to display the equation and R^2 values on the plot.

Table 1.1

Dilution	BSA Protein Concentration	Absorbance at 595 nm
1 (concentrated)	2 mg/mL	
1:2		
1:4		
1:8		
Blank	0 mg/mL	

STEP THREE

Work in groups of four to use the Bradford Assay and your standard curve to determine the concentration of protein in an unknown solution provided by the teaching assistant for your laboratory section.

1. Obtain a 1.5-mL microfuge tube containing a protein of unknown concentration from your teaching assistant. Make note of the letter written on the tube.

2. Set your p100 micropipette to 50 µL (equivalent to 0.05 mL).

3. Attach a yellow tip to the barrel and draw up 100 µL of the solution. Dispense into fresh cuvette.

4. Add 1500 µL of the Bradford Assay reagent (Coomassie Brilliant Blue G-250 dye) to the microfuge tube that contains 50 µL of your solution with an unknown protein concentration.

5. Snap the cap on the cuvette and invert it to mix it well.

6. Let the cuvette sit at room temperature for five minutes.

7. Make sure the spectrophotometer is set to 595 nm absorbance. Insert the cuvette and record the absorbance at 595 nm.

8. If time permits, perform points 1 through 7 for a second and even a third time so as to confirm your result.

9. Use the standard curve you generated in Step Two and the absorbance value(s) you obtained for your unknown to determine the concentration of protein in the solution provided by the teaching assistant for your laboratory section.

CLEAN-UP

1. Pour the cuvette contents down a drain. Rinse each three times, then place on a paper towel for drying.

2. Close Eppendorf microcentrifuge tubes and place in trash.

3. Return all items from your bench into bin.

4. Clean bench with 10% bleach solution and paper towels.

5. Wash hands before exiting lab.

IMPORTANT QUESTIONS TO CONSIDER

Careful consideration of the following questions related to this exercise will not only insure that you understand the most important concepts of this laboratory but will also help you prepare for the lecture portion of the course. By addressing these questions (and others that you decide are important) in the conclusions section of your laboratory report, you will also assure that you receive full credit for that portion of your report.

- PLEASE PROVIDE A completed version of Table 1.1.

- PLEASE PROVIDE THE standard curve you generated for five solutions with different concentrations of BSA.

- WHAT WAS THE equation for the best fit linear regression of your standard curve?

- WHAT WAS THE correlation coefficient (r-squared value) for your standard curve?

- OVER WHAT RANGE of protein concentrations would you be able to use your standard curve to determine the concentration of a protein in an unknown?

- IF YOU WERE able to generate two sets of data points for your standard curve, how similar were they? What might account for differences between the two replicates?

- WHAT WAS THE letter on the tube provided by your teaching assistant and what was the concentration of protein in that solution?

EXPERIMENT 2:
CONCENTRATION, *p*H, AND BUFFERS

OVERVIEW

Biology is the study of living systems, and biological scientists observe living things and their component parts. In scientific investigation these observations then become the foundation of interesting questions that require experiments and further observations. An underlying assumption that scientists make in this process is that biological systems are understandable and can be explained by fundamental rules or laws.

The same fundamental rules or laws that apply to biological systems also apply to other areas of scientific investigation. In fact, many prominent biologists have received their primary training as scientists in such varied disciplines as chemistry, physics, and mathematics.

Of central importance to virtually all work in biological sciences is the concept of the concentration of solutes and hydrogen ions—areas that are usually thought to fall more into the realm of chemistry than biology. Both cell structure and function can be radically altered by extremely small fluctuations in the concentrations of solutes and hydrogen atoms. During the course of this experiment, you will use several different ways that can be used to determine the concentration of hydrogen ions. At the same time you will also study how their concentration in solutions can be changed or maintained by living systems.

INTRODUCTION

Solutions are homogeneous mixtures. They are usually classified according to their physical state; gaseous, liquid, and even solid solutions can be prepared. Liquid solutions are the most common and are usually of greatest interest to biologists. Most of the theories central to our understanding of liquid solutions were first put forward in the 1870s and 1880s by chemists such as Raoult, van't Hoff, Arrhenius, and Gibbs whose names you will encounter in the names of formulae and variable's in chemical equations.

The component of a solution that is present in the greatest quantity is usually called the solvent. In virtually all biological systems that **solvent** is water. All components other than the solvent in a solution are called **solutes.**

The concentration of a solute in a solution can be described in several different ways. Biologists usually use **molarity** to describe the concentration of solutes in solvents though "percentage by mass," "molality," "normality," and "mole fraction" are alternatives that are often used in other disciplines. The molarity, M, of a solution is the number of moles of solute per liter of *solution* (not solvent—the difference can be very important). Thus, 1 liter of a 1 M solution of sodium chloride (table salt) is prepared by taking 1 mole (6.022×10^{23} pairs of sodium and chlorine atoms which is equivalent to the molecular weight in grams, 58.5 grams, of salt) and adding sufficient water to make exactly 1 liter of solution. [Most people make their coffee a roughly 1 M solution of sucrose (table sugar) before drinking it by adding a teaspoon of sugar per cup.]

Pure water itself is actually a very weak **electrolyte** because it can dissociate into two charged molecules. When a water molecule dissociates (or ionizes), a hydrogen atom that is covalently bound to the oxygen of a water molecule leaves its electron behind and, as a hydrogen ion (H^+), joins a different water molecule. Two ions, a hydroxide ion (OH^-) and a hydronium ion (H_3O^+), are produced by this reaction as shown in Equation (2.1).

$$2\,H_2O \longleftrightarrow H_3O^+ + OH^- \tag{2.1}$$

In any amount of pure water or in any solution, a small number of water molecules are dissociated and exist in this ionized state. As Equation (2.1) indicates, the number of hydronium ions in pure water always equals the number of hydroxide ions since one cannot be formed without the other also being made. In pure water the concentration of hydronium ions, $[H^+]$, is 1×10^{-7} M and the concentration of hydroxide ions, $[OH^-]$, is also 1×10^{-7} M. Thus, any liter of pure water is actually a 0.0000001 M solution of hydronium and hydroxide ions.

By convention, mathematicians use the letter "*p*" to represent the negative logarithm of a number (the negative of the power to which 10 must be raised in order to get the number). Thus the hydronium ion concentration of water can be conveniently expressed as its $p[H^+]$ or, more simply, as its *p*H. The *p*H of pure water for instance is 7 (the negative logarithm of 1×10^{-7} M). The concentration of hydroxide ions can be expressed similarly as *p*OH.

The *p*H and *p*OH of a solution need not always be equal. For instance, some solutes like hydrochloric acid (HCl) bring an excess of hydrogen ions with them and increase their concentration in water. Since all the hydrogen ions in 0.01 mole of hydrochloric acid will dissociate from their chloride ion counterparts when they are dissolved in 1 liter of water, a 0.01 *M* solution of hydrochloric acid will actually have a hydronium ion concentration of 0.01 *M* plus 0.0000001 *M* of the water itself or 0.0100001 M. For most experiments it is safe to consider 0.0100001 *M* to be essentially the same as 0.01 *M* and to say that a 0.01 *M* solution of hydrochloric acid is *p*H 2.

While *p*H and *p*OH of a solution need not always be equal, the chemistry of water actually requires that their sum must always equal 14. Thus, a solution with a *p*H of 2 must have a *p*OH of 12 (in this case *p*OH = 14 − 2). By definition, a solution that has the same *p*H and *p*OH (7) is said to be neutral. A solution with a *p*H that is less than its *p*OH (meaning that the *p*H is less than 7 as in the case of the 0.01 *M* hydrochloric acid solution in the previous paragraph) is said to be acidic. A solution with a *p*H that is greater than its *p*OH (meaning that the *p*H is greater than 7) is said to be basic or alkaline. Substances that increase the concentration of hydronium ions (decrease *p*H) are known as **acids** while those that increase the concentration of hydroxide ions (raise *p*H) are called **bases.**

It is important to remember that the "*p*H scale," like the Richter scale for measuring the strength of earthquakes, relates to powers of 10. A solution with a *p*H of 1 has a hydronium ion concentration that is *100* times more acidic than a solution with a *p*H of 3.

Living systems are extremely sensitive to *p*H and generally function best when conditions are nearly neutral. The blood of most mammals, for instance, is usually maintained in the *p*H range of 7.35 to 7.45. Very small changes in *p*H can have dramatic effects upon the structure and function of cells and the molecules from which they are built. For instance, one class of molecules called **anthocyanins** that are responsible for many of the dark reds, blues, and purples of flowers, fruits, and autumn leaves have their characteristic colors only within limited *p*H ranges. Roses, strawberries, and cornflowers use an anthocyanin with precisely the same chemical structure though in the neutral *p*H found in the cells of rose flowers and ripe strawberries it is red in color. The very same anthocyanin in the basic (high *p*H) conditions within the cells of cornflowers however has a pale blue color. As part of this experiment, you will be using an anthocyanin extracted from cabbage leaves that has an intense red color in some *p*Hs but is colorless in other *p*Hs (Figure 2.1).

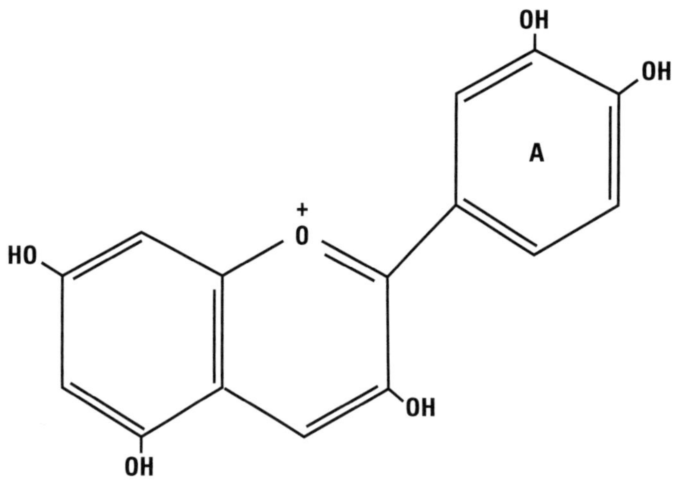

Figure 2.1 Cyanidin: the purplish red anthocyanin of red cabbage. The addition of a single hydroxyl (OH) group to the ring labeled "A" would convert cyanidin to the bluish purple delphinidin of morning glory flowers while the removal of one would make it pelargodnidin which is responsible for the orange-red color of ripe pumpkins.

Departures from a given cell's characteristic *p*H can have dramatic (and devastating) consequences for an organism. As a result, living systems employ a wide variety of mechanisms to regulate their internal *p*H. The most powerful and simplest of those mechanisms are **buffers.** Buffered systems contain solutes that allow them to maintain their *p*H at a fairly constant value even when small amounts of acid or base are added. The solutes in buffers are themselves either acids or bases that do not completely dissociate into their ionic forms. These "weak" acids or bases essentially act as reservoirs that can absorb or release limited amounts of hydrogen or hydroxide ions contributed by "strong," fully dissociating acids or bases that are added to the solution. The primary buffer system within our own blood relies upon differing ratios of carbonic acid (H_2CO_3) and bicarbonate (HCO_3^-) and it helps keep our extracellular fluids within a very narrow *p*H range of between 7.34 and 7.42.

PROCEDURES

Before coming to the laboratory, become more familiar with the terminology of pH by completing Table 2.1. Fill in all the missing hydronium and hydroxide ion concentrations, pH and pOHs, and label the solutions as either acidic, basic, or neutral. Include the completed table in the results section of your laboratory report.

Table 2.1 The pH scale.

[H$^+$] M	pH	[OH$^-$]	pOH	Type of Solution
10^0 (1.0)	0	10^{-14}	14	Acid
10^{-1}	1	10^{-13}	13	
10^{-13}				
10^{-2}		10^{-12}		
10^{-4}		10^{-10}		
	8		6	
	3		11	Acid
10^{-5}		10^{-9}	9	
10^{-6}		10^{-8}		
		10^{-7}		Neutral
			5	
	10	10^{-4}	4	
10^{-11}		10^{-3}	3	
10^{-12}		10^{-2}		
10^{-14}		10^0 (1.0)	0	Basic

Prior to your arrival in the laboratory, the prep-room personnel will have prepared an anthocyanin solution from red cabbage for you to use as an indicator of the *p*H of a wide variety of solutions. Fresh red cabbage leaves will be boiled in water for three minutes and filtered through cheesecloth so that only the purple-red anthocyanin (cyanidin) will remain in the solution. Dyes prepared in very similar ways have been used by many cultures for thousands of years to provide colorful pigments for clothing.

To use the anthocyanin in this solution as a *p*H indicator, you must first determine how its color changes when it is exposed to other solutions whose *p*H is already known. **Standards** of known *p*H ranging from 2 to 14 will also be provided. By comparing the color of the anthocyanin when mixed with these standards of known *p*H with its color when mixed with solutions of unknown *p*H, it will be possible for you to determine the *p*H of the solutions you test.

STEP ONE

1. Work in pairs. Obtain eight clean test tubes and a rack from your teaching assistant and label the tubes "2," "4," "6," "7," "8," "10," "12," and "14." Use re-pipettors to deliver 5 mL of the eight solutions of known pH to the tube whose number corresponds to the solution's pH.

2. Use a pipette to transfer 3 mL of the cabbage extract to each of the eight labeled tubes. Keep in mind that while the purple-red color of cyanidin may be attractive in the tubes, you might find its color less appealing if you stain your clothes with it. Carefully swirl each tube to mix their contents thoroughly.

3. Place a white piece of paper behind each tube and observe the color of the mixed solutions. Record the colors you observe in each tube in Table 2.2 below. Note any changes in color that occur gradually as well and include Table 2.2 in the results section of your laboratory report.

Table 2.2 The color of red cabbage anthocyanin in solutions of known pH.

pH of Standard	Color of Anthocyanin	Observations
2		
4		
6		
7		
8		
10		
12		
14		

STEP TWO

1. Continue to work in pairs. Obtain six additional clean test tubes from your teaching assistant and label the tubes "A," "B," "C," "D," "E," and "F."

2. Take the tubes labeled A through F to the side of the laboratory closest to the prep-room and the door and choose six solutions from those provided there. Record the solutions you have chosen in Table 2.3. Use the pipettes provided near the six solutions you have chosen to transfer 5 mL of each of these solutions into its own labeled test tube using the key you created in Table 2.3.

3. Return to your laboratory bench and add 3 mL of the cabbage extract indicator to each of the tubes labeled "A" through "F." Gently swirl each tube to mix their contents thoroughly.

4. Place a white piece of paper behind each tube and observe the color of the mixed solutions. Directly compare the colors of these tubes to the colors of your eight standards prepared in the previous step. Record the colors you observe in each tube and the *p*H they correspond to in Table 2.3.

Table 2.3 The *p*H of six household solutions using an anthocyanin indicator.

Solution	Name	Color of Anthocyanin	Estimated *p*H
A			
B			
C			
D			
E			
F			

STEP THREE

The cabbage extract used in the previous two steps is most useful for determining the pH of solutions with little or no color of their own. Alkacid test strips are pieces of paper that have been impregnated with several indicators whose colors also change with pH and often prove more useful in determining the pH of colored solutions.

1. Obtain six pieces of alkacid test paper from your teaching assistant. Return to your laboratory bench.

2. Dip the end of one of the pieces of alkacid paper into each of the six tubes and record the color of the moistened paper in Table 2.4. Compare the color of the moistened paper to the color standard provided for alkacid test paper and record the corresponding pH in Table 2.4 as well.

Table 2.4 The pH of six household solutions using alkacid test paper.

Solution	Name	Color of Alkacid Paper	pH
A			
B			
C			
D			
E			
F			

STEP FOUR

The *p*H estimates obtained with anthocyanin indicators and alkacid test papers often differ due to the subjectivity involved in visually comparing sample colors to standards. More accurate and objective measures of *p*H are often obtained in biological sciences laboratories through the use of electronic *p*H meters. These more precise meters consist of a glass electrode that is sensitive to hydrogen ion concentration gradients and a reference electrode that completes an electrical circuit.

Remember that buffered solutions are ones that contain a solute that causes it to resist changes in *p*H when small amounts of acid or base are added to it. A mixture of a weak acid (such as phosphate, $H_2PO_4^-$) and its salt (such as potassium phosphate, KH_2PO_4) can both give up hydrogen ions (from the phosphate which acts as an acid) and take them up (by the displacement of potassium from potassium phosphate which acts as a base).

Phenolphthalein is another chemical compound whose color dramatically changes in a *p*H dependent fashion. However, unlike the other indicators you have used in the previous steps of this experiment, phenolphthalein undergoes only a single dramatic change in color (either to or from being colorless or intensely pink) at a single, well-defined *p*H.

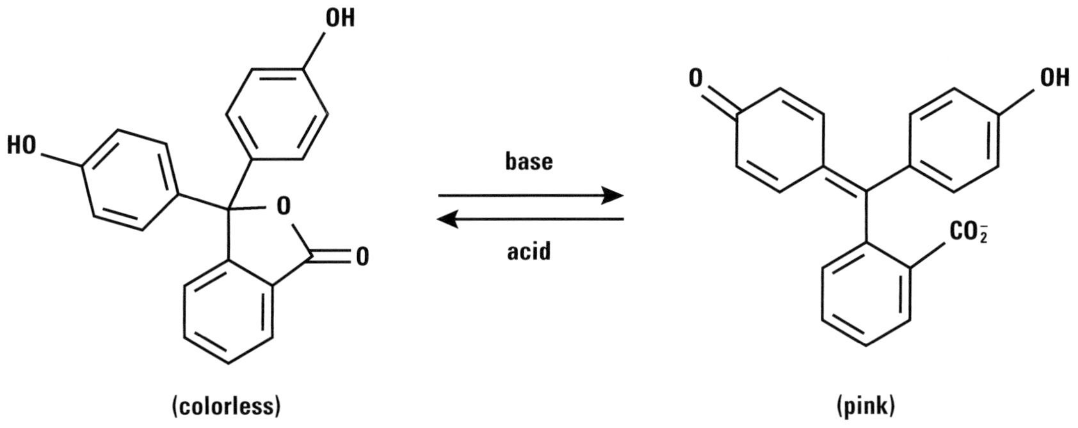

Figure 2.2 Phenolphthalien: The chemical structure cannot absorb visible light in acidic solutions but does absorb all but pink light in basic solutions.

In this step of the experiment, you will observe your teaching assistant gradually add a dilute hydrochloric acid solution to a solution that is not buffered and one that contains a phosphate buffer. By recording the pH of the solutions as determined with a pH meter and by watching the amount of base needed to cause a transition in phenolphthalein's color, the ability of buffers to moderate changes in pH should be apparent.

1. In a small group of six to eight students, observe your teaching assistant add 50 mL of a solution that is not buffered into a 100-mL beaker on a magnetic stirrer. To this solution a pipettor will be used to transfer 1 mL of a dilute solution containing phenolphthalein.

2. Record the color and pH (as determined with an electronic pH meter) of the solution in Table 2.5.

3. Observe as your teaching assistant adds three drops of dilute sodium hydroxide to the solution and record the pH and color of the solution in Table 2.5.

4. Repeat the previous process (3) until the color of the solution changes permanently.

5. Note that the addition of a strong acid such as hydrochloric acid (HCl) is sufficient to quickly return to a solution to its original color due to its effect on the solution's pH.

6. Observe your teaching assistant replace the 50 mL of un-buffered solution with 50 mL of a solution containing a phosphate buffer. To this solution, 1 mL of a dilute solution containing phenolphthalein will once again be added.

7. Repeat points 2 through 5 in this step.

Table 2.5 Titration of an un-buffered and a phosphate-buffered solution.

Number of Drops Added	Color of Phenolphthalein	*p*H of Solution

8. Use the grid provided below to plot the pH of the un-buffered and phosphate-buffered solutions as drops of base were added. Use an arrow to indicate the point on the curves at which a change in the color of phenolphthalein was observed. Include this graph as part of your lab report.

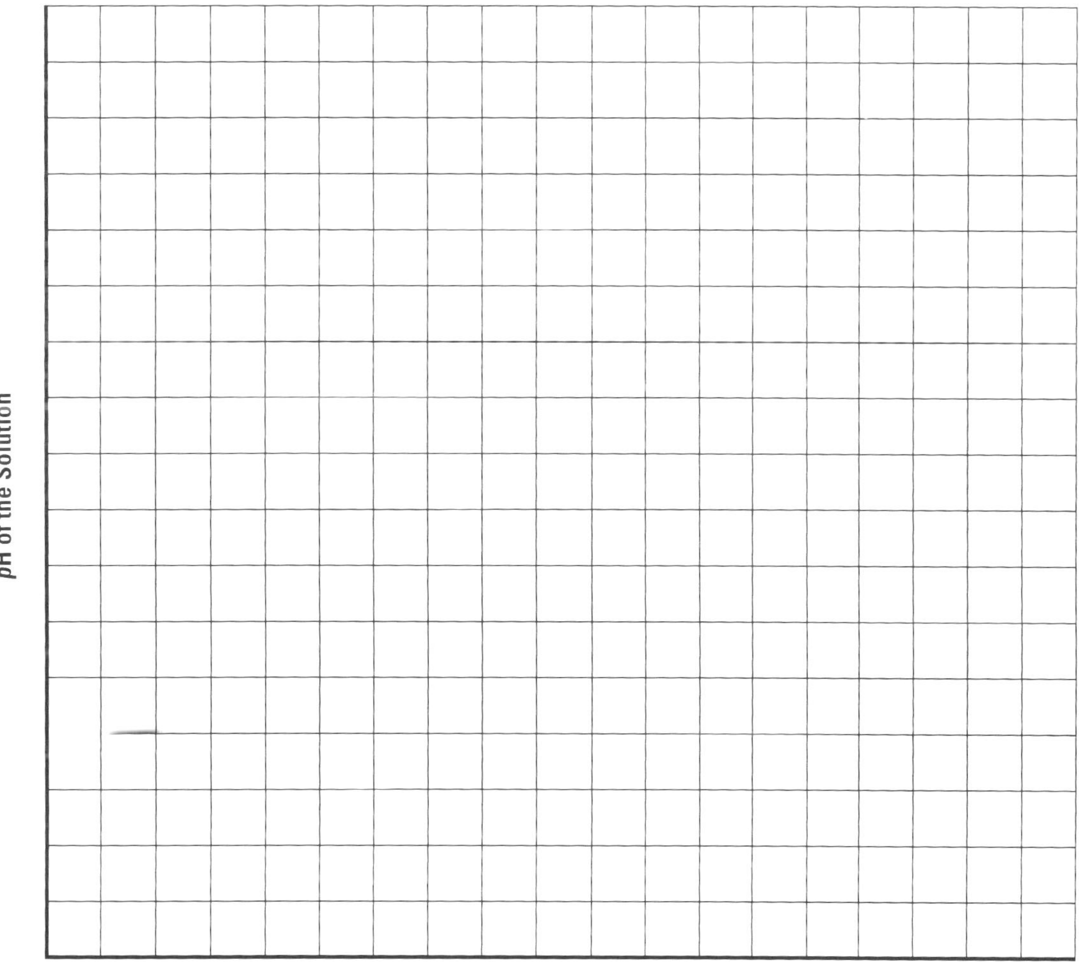

IMPORTANT QUESTIONS TO CONSIDER

Careful consideration of the following questions related to this set of experiments will not only insure that you understand the most important concepts of this laboratory but will also help you prepare for the lecture portion of the course. By addressing these questions (and others that you decide are important) in the conclusions section of your laboratory report, you will also assure that you receive full credit for that portion of your report.

- WHAT IS THE molar concentration of hydrogen ions within a healthy human's blood?

- OVER WHAT pH range would the anthocyanin in the red cabbage extract be an effective indicator of the pH of a solution?

- HOW WIDELY DID the pHs of the household solutions you used differ?

- DID THE CABBAGE extract indicator and the alkacid test paper you used report the same pHs for the household solutions you examined? Give a reason why they might have differed and which do you think is more reliable?

- AT WHAT pH did phenolphthalein's color change and in what situations would it make a useful indicator?

- HOW DID THE un-buffered and phosphate-buffered solutions differ in their response to the addition of sodium hydroxide (NaOH) and what role might buffers play in living systems?

- INCLUDE COMPLETED versions of Tables 2.1, 2.2, 2.3, 2.4, 2.5, and your titration curve with your answers to this week's questions.

EXPERIMENT 3:
OSMOSIS AND DIFFUSION

"The organization of water within biological compartments is fundamental
to life, and the aquaporins serve as the plumbing systems for cells."
Peter Agre (Nobel Prize winner for the discovery of auaporins)

OVERVIEW

All molecules tend to move from regions of high concentration to regions where they
are in lower concentration. Movement down such concentration gradients is known as
diffusion. When the molecule moving down a concentration gradient is water, osmosis
is said to have occurred.

Because water is such a good solvent and would normally bring other molecules
along with it, osmosis requires the presence of a semi-permeable membrane. As their
name implies, semipermeable membranes allow the passage of some molecules (often
relatively small ones like water) but not others (often relatively large ones like starches,
proteins, and DNA).

In this experiment you will use a semipermeable membrane to determine the relative
sizes of atoms, water and a complex carbohydrate (starch). You will also use the
semipermeable nature of the membranes of living cells to determine the concentration
of relatively large molecules inside of those cells.

INTRODUCTION

Osmosis is a classic example of passive transport in that no energy needs to be provided for the net movement of water molecules to occur. Living cells use active transport all the time to change and/or maintain the concentration of many different molecules inside of them relative to their concentration outside of them. This typically results in cells accumulating high concentrations of molecules that they need (like sugars and amino acids) inside of themselves and high concentrations of things that they do not need (like waste materials) outside of themselves. When chemical energy such as that associated with adenosine triphosphate (ATP) is used, primary active transport is said to have occurred. In contrast, if the movement is driven by the establishment of an electrochemical gradient then secondary active transport has occurred.

The phospholipid bilayer of cell membranes makes an effective barrier to the movement of water. Water molecules with their slightly negatively charged oxygen atoms and their slightly positively charged hydrogen atoms make for a very versatile solvent. But, those slight charges make it very difficult for water molecules to pass through the hydrophobic interior of a phospholipid bilayer. The transport of water and other molecules that have charges associated with them through cell membranes is typically accomplished by having the molecules pass through specialized protein pores that are embedded in cell membranes. Most of those protein pores (such as the aquaporins used for water transport or lactose permease which allows the passage of the sugar lactose) are very specialized and will only allow one specific kind of molecular to move across a cell's membrane.

It is often useful to talk about concentration in relative terms. When one solution has a lower concentration of solutes than another solution, the first solution is said to be **hypotonic** relative to the second solution. In the same way, a solution with a higher concentration of solutes relative to another solution is said to be **hypertonic.** When two solutions have the same concentration of solutes, they are said to be **isotonic.** The combination of the strength of osmosis and the semipermeable nature of the membranes of cells can have dramatic effects on living tissues at a cellular level. Consider the human red blood cells portrayed in Figure 3.1. Red blood cells in a hypertonic solution appear shriveled and prune-like while those in a hypotonic solution swell and can even burst open.

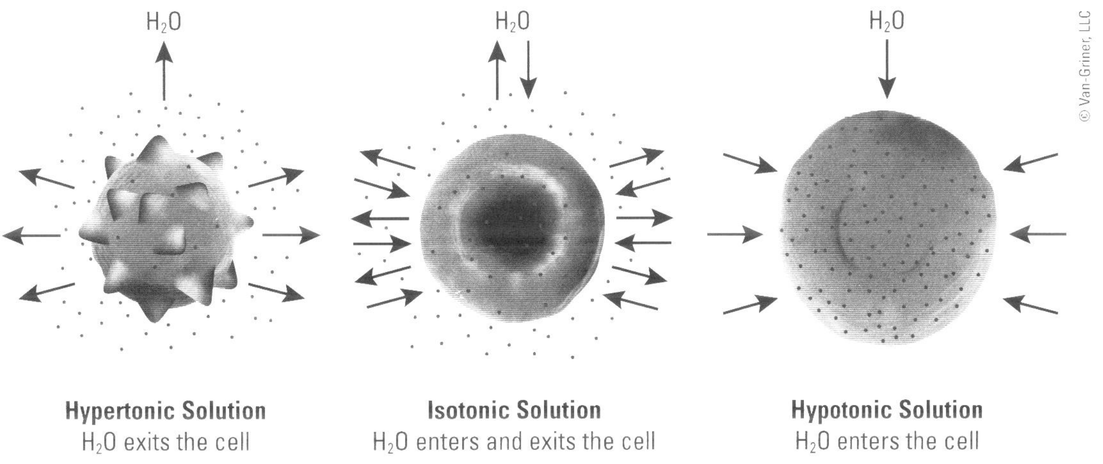

Figure 3.1 Human red blood cells in solutions with different concentrations of solutes. A solution that is 5% glucose would be isotonic to red blood cells.

As you might imagine, a certain amount of force needs to be applied to break open a cell. That same force can be used to raise a column of water in a U-shaped tube where a semipermeable membrane at the bottom of the U separates a watery solution containing a solute like a sugar from one that is just pure water. If the sugar molecules are too large to pass through the semipermeable membrane's pores but water molecule are not, then there will be a net movement of water molecules from the pure water side of the tube to the side that contains the sugar and cause the level of the solution on that side of the tube to be higher than that on the pure water side. Movement of water in such a U-tube will stop only when the weight of the column of solution on the sugary side counteracts the osmotic force. This principle is commonly used to desalinate ocean water (sea water can be placed in a chamber then put under pressure that is greater than the sea water's osmotic pressure to force purified water across a semipermeable membrane). Osmotic pressure is also essential for many important functions in plants such as the turgor pressure on cell walls that allow herbaceous plants to stand upright.

PROCEDURES

STEP ONE

Dialysis tubing are actually semipermeable membranes. Some have pore sizes that are too small to allow molecules like starches and even water or oxygen (O_2) gas to pass through but not too small to allow atoms to diffuse through them. You will follow the movement of iodine atoms across the membrane of some dialysis tubing by taking advantage of the fact that starch turns a dark purple color when exposed to potassium iodide (I_2KI).

1. Work in pairs. Cut a six inch-long section of dialysis tubing from the roll of tubing. Soak your piece of dialysis tubing in water for one minute to separate its layers. After soaking, use a dialysis clip to close one end of the dialysis tubing. Use a pipette to transfer 20 mL of cornstarch solution to the dialysis tubing and seal with another dialysis clip. Rinse tubing with distilled water after filling and sealing.

2. Add 300 mL of water to a 500-mL beaker along with a magnetic stir bar. On a magnetic stir plate, add drops of potassium iodide solution until solution is golden brown. (Potassium iodide will stain your skin and your clothing if it contacts them.)

3. Place the dialysis tubing containing the cornstarch solution into the beaker so that the cornstarch mixture is submerged in the potassium iodide solution.

4. Make notes regarding the change in the color of the potassium iodide solution in the beaker and the cornstarch solution in the dialysis tubing over the course of the remainder of the laboratory period.

5. Discard the dialysis tubing containing the cornstarch solution. Discard the potassium iodide solution in a sink. Rinse and return the beaker and the dialysis clips to your teaching assistant.

In the course of the second set of experiments in this laboratory, you will investigate the concentration of molecules like sugars, starches, amino acids, and proteins in living potato cells that are too large to pass through the semipermeable membrane of the phospholipid bilayer of potato cells.

STEP TWO

1. Work in pairs. Obtain seven large, clean test tubes and a rack from your teaching assistant and label the reaction tubes "0," "0.2," "0.3," "0.4," "0.5," "0.6," and "0.8." Use re-pipettors to deliver 5 mL of the seven solutions with different molar concentrations of glucose (ranging from 0 to 0.8 M) that will have been prepared by the prep-room personnel to the tube whose number corresponds to the solution's molar concentration of glucose.

2. Use a cork borer to make seven cylinders of potato. Push the borer through the cut side of the sliced potatoes that have been cut into approximately 3 cm long pieces. Avoid including the potato skin (edges) in the potato cylinders that you make. Gentle twisting back and forth will help ease the cork borer through the potato slices. You will need seven complete cylinders of potato of approximately the same length before you can proceed to the next step of this experiment.

3. Use a paper towel to gently remove excess moisture from the sides and ends of the potato samples. Place all seven potato samples in a Petri dish.

4. Weigh one of the potato samples to the nearest 0.01 gram on the aluminum sheet on the laboratory balance. Record the weight in Table 3.1 of this laboratory manual, use a razor blade to slice the cylinder in half, then add both halves to the test tube labeled "0" and record the time in Table 3.1.

5. Repeat point 4 for potato samples that will be placed in the test tubes labeled "0.2," "0.3," "0.4," "0.5," "0.6," and "0.8."

6. Incubate all seven potato samples for approximately 45 minutes, gently swirling each tube every 10 to 15 minutes. (This would be a good time to check on the beaker you set up for the Step One portion of this laboratory.)

7. Remove the potato pieces from the test tube labeled "0" and use a paper towel to gently remove excess moisture (but don't squeeze the potato pieces). Weigh the potato pieces and record their final weight in Table 3.1. Discard the potato sample.

8. Repeat point 7 for the potato samples that were placed in the test tubes labeled "0.2," "0.3," "0.4," "0.5," "0.6," and "0.8" using the same order you used in point 4. Empty the cornstarch solution into a sink and discard the plastic sandwich bag. Discard the potassium iodide solution in a sink. Rinse and return the beaker to your teaching assistant.

9. Rinse and return the test tubes you have been using to your teaching assistant. After leaving the laboratory, use the data you recorded in Table 3.1 to construct a plot of percent change in potato sample weight *versus* the molar concentration of the glucose solutions you used. Be certain to include the plot in the results section of your laboratory report.

Table 3.1 The weights of five potato samples exposed to solutions with different molar concentrations of glucose.

Glucose (M)	Initial Weight (g)/Time	Final Weight/Time	Percent Change
0			
0.2			
0.3			
0.4			
0.5			
0.6			
0.8			

IMPORTANT QUESTIONS TO CONSIDER

Careful consideration of the following questions related to this set of experiments will not only insure that you understand the most important concepts of this laboratory but will also help you prepare for the lecture portion of the course. By addressing these questions (and others that you decide are important) in the conclusions section of your laboratory report, you will also assure that you receive full credit for that portion of your report.

- FOR WHICH OF the five solutions with different concentrations of glucose were your potato samples isotonic? Which were hypertonic and which were hypotonic?

- TURN IN A completed version of Table 3.1. Use the information in the completed table to construct a graph that shows the percent change in the final weight of the potato samples relative to the molar concentration of glucose in the solutions you used.

- WOULD INCREASING THE temperature of the glucose solutions that you used by 5°C have caused your conclusions to be different?

- YOU USED PLANT cells in this experiment. Predict what might have happened if you had used cylinders of animal cells (e.g., chicken muscle cells from chicken breast) instead of potato cells.

- AN INTRAVENOUS ADMINISTRATION of 5% glucose (dextrose) is commonly used as a maintenance fluid for patients unable to drink water. Do you expect that solution to be hypertonic, isotonic, or hypotonic relative to red blood cells?

- BASED ON YOUR observations, which substance moved across the membrane of the dialysis tubing, the potassium iodide or the starch?

- WHAT ARE THE sizes of the pores in the plastic used to make dialysis tubing (e.g., are they large enough for atoms, water, sugars, or starches to pass through)? To what other things might the dialysis tubing you used be semipermeable?

EXPERIMENT 4: MICROSCOPES AND CELLULAR ANATOMY

"People who look for the first time through a microscope say 'Now I see this, and then I see that,' and even a skilled observer can be fooled. On these observations I have spent more time than many will believe, but I have done them with joy." Antonie van Leeuwenhoek

OVERVIEW

The microscope has become one of the most recognizable symbols of science. Virtually all of our understanding of subcellular structure has been discovered with or confirmed by the development of better and more powerful microscopes. These insights have become the cornerstones of contemporary thinking in such diverse areas as evolution, biotechnology, pharmacology, and medicine.

Just as an understanding of molecular interactions is fundamentally important to understanding the working of cells, an appreciation of cells is essential for a thorough understanding of organisms as a whole. Microscopic examination of individual cells reveals a remarkable unity and diversity in living things.

During the course of this experiment, you will be introduced to the operation of high-powered microscopes. With these valuable tools you will not only have an opportunity to explore a complex and fascinating new world but also examine first-hand the anatomical features of cells that are common to all living things and gain insights into the roles those features play.

INTRODUCTION

At its very best, the unaided human eye has a **resolving power** of about 0.1 millimeter (~100 µm) meaning that we can recognize two individual points as separate only if they are at least that far apart. We perceive any pair of points closer together than that distance as a single fuzzy point if at all. It is only with the magnifying help of curved optical lenses that we can resolve smaller objects and begin to study individual cells whose diameters typically range from between 0.0005 and 0.05 mm (0.5 and 50 µm).

The laws of optical physics and the magnifying properties of curved surfaces have been widely appreciated by philosophers and physicians since at least 300 BC. Strangely though, it has only been within the last 400 years that scientists have used magnification as an important tool in their investigations related to such diverse fields as astronomy, paleontology, chemistry, and biology.

One very early and isolated reference to the use of a lens for the purpose of magnification comes from Pliny the Elder who wrote between 23 and 79 AD that: "Emeralds are usually concave so that they may concentrate the visual rays. Emperor Nero used to watch in an emerald the gladiatorial combats." This ancient historian appears to have written the first description of a monocle used for correcting short-sighted vision.

Nero's emerald eye-piece is likely to have been very effective yet, even though many people must have read Pliny the Elder's history, there is presently no other evidence of spectacle use until after the passage of twelve more centuries. During the Renaissance in Florence, Italy, however, the use of spectacles finally became widespread and immensely popular. Nonetheless, it still took approximately 350 more years still for telescopes and, shortly thereafter microscopes, to be constructed using the same fundamental principles of refraction.

Many people mistakenly believe that Antoni van Leeuwenhoek invented the microscope. He actually only began using simple and crude microscopes of his own design sixty to seventy years *after* very elaborate models were commercially available in Europe. True credit for the first microscope appears to belong to Zacharias Jansen who began manufacturing and selling them in Holland in 1595. While many important discoveries had already been made using microscopes of Jansen's design, it was van Leeuwenhoek who published some of the first microscopic studies of living things for which he is still famous.

Even still, it was not until Robert Hooke published "*Micrographia*" in 1655 that the utility of microscopy in biological studies became widely appreciated. The image in his work that has left the most enduring mark on biology was that of a thin slice of cork. From that image it became clear that cork was a mesh-work of a material making a supporting structure around tiny pockets of air. These pockets reminded Hooke of the tiny cells in which monks of his time lived and he named them accordingly. While he did not appreciate that the air pockets he saw were actually the hollow remains of what are now considered to be cells, the name he chose for them persists.

It was not until 1809 that the French naturalist Jean Baptiste de Lemarck put forward the idea that all living things are composed of cells and that cells were the smallest discrete unit that could be considered to display all the properties associated with life. By the mid-1800s this **cell theory** was well established and in 1858 the German physician Rudolf Virchow proposed an important extension—that all living cells arise from pre-existing cells ("*omnis cellula e cellula*"). It was Louis Pasteur, working in France, who a few years later supplied definitive proof of Virchow's ideas of **biogenesis.**

Studies in the last 100 years have revealed a wealth of additional information about the organization and function of cells. While no one cell type can be considered "typical," most cells are too small to be seen without the aid of a microscope. The smallest cells (0.5–5 μm in diameter) are those of **prokaryotes**—relatively simple, single cell organisms that lack a true nucleus and membrane-bound organelles. The more complex cells of **eukaryotes** (typically 5–50 μm in diameter) are not only larger than prokaryotic cells but also have a nucleus as well as other membrane-bound organelles such as chloroplasts and mitochondria.

Some eukaryotic organisms live as single cells just as prokaryotes do. Most eukaryotes though are aggregates of large numbers of cells that all share the same genetic material but are extensively specialized to perform specific tasks. For instance, eukaryotes such as ourselves are made of billions of cells and thousands of different cell types. Microscopes allow rapid confirmation of cell types based upon which distinguishing and functionally important assortments of microscopic features they are found to possess including organelles and structures such as: nuclei, mitochondria, chloroplasts, cell walls, vacuoles, and cilia.

PROCEDURES

Prior to your arrival in the laboratory, you should become familiar with the proper use and care of the compound microscopes you will be using. Refer to Figure 4.1, a diagram of a compound microscope like the ones that will be used in your laboratory section, as you read the following introductory material.

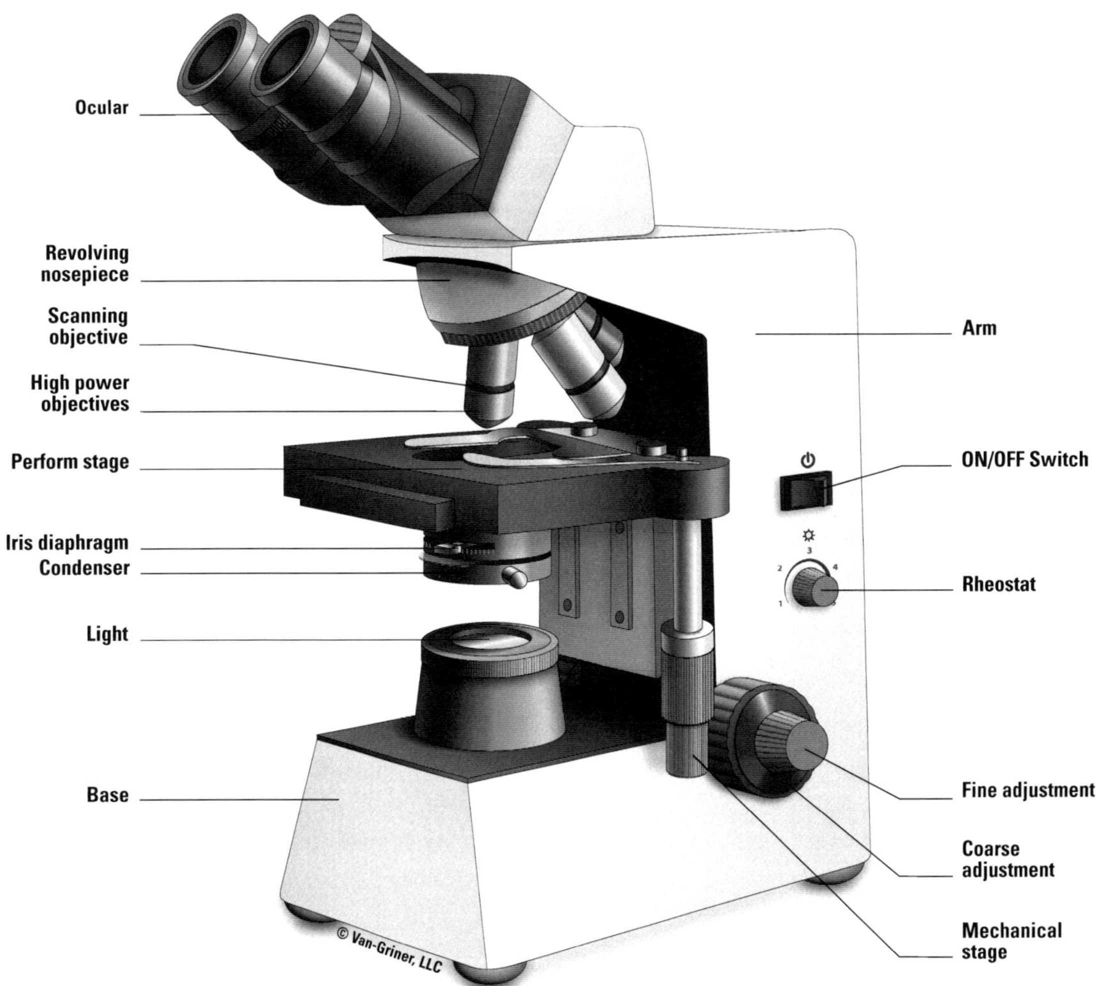

Figure 4.1 Compound microscope.

INTRODUCTION TO THE COMPOUND MICROSCOPE

- **Mechanical parts:** The mechanical, non-optical components of the microscope serve the dual purpose of supporting the optical components and making the microscope easy to use. To function properly, they must be very precisely aligned and move with precision. Even though they have been heavily constructed, they should always move very easily—*never* force them.

  The **base, arm,** and **body tube** (to which the eyepiece is attached) are the main elements for supporting the optical systems of a compound microscope. A **nosepiece** such as the one illustrated in Figure 4.1 can hold multiple lenses and allows rotation of the desired **objective** lens into the light path. A **stage** with **clips** are used to hold samples firmly in place and is usually accompanied by a micrometer that allows specimens to be moved precisely both from side to side and forward and backward. **Coarse** and **fine adjustment** knobs are used to bring objects into focus by moving the stage up and down and consequently change the distance between specimens and an objective lens.

- **Optical parts:** The illumination needed to examine specimens microscopically is provided by a system composed of: a **light source** (usually a small light bulb with an intensity control), an **iris diaphragm** (which can be opened and closed to increase or decrease the amount of light passing to a specimen from the light source), and the **condenser** lens that focuses the beam of light from the light source on the specimen.

  A separate imaging system consists of an additional set of two lenses. The **objective** lens is the one nearest the object being examined while the **ocular** lens (eyepiece) is the one used to actually view specimens. Both of these lenses are actually composed of several optical parts and must be used together to magnify images and provide resolution so that objects many times smaller than the 0.1-millimeter limit of the human eye can be seen.

- **Magnification:** The total magnification of a specimen's image when viewed through a compound microscope is the product of the individual magnification of each of the two sets of lenses used (the objective and the ocular). Standard ocular lenses magnify images ten-fold (10×) and the magnifying power of your microscope's ocular will be marked on its surface. The objective lenses located on the revolving nosepiece of your microscope will each have a different power: 4×, 10×, and a high-power 40×. Thus, in conjunction with the ocular lens, the objectives of the microscope you will use allow between 40- and 400-fold magnification of the actual size of the objects you view.

  In addition to the compound microscopes you will be using in this set of experiments, a special microscope with an **oil immersion** objective will be set up as a demonstration for you. The refractive properties of oil relative to air allow the use of a 100× objective and yield a magnification of 1,000×—very near the usable limit of image magnification with light microscopy.

- **Field of view, depth of field and light intensity:** The laws of optical physics are such that no single objective lens is best for all purposes. The low power 4× objective lens is the easiest to use to find and focus on objects on the stage because it takes in a wider field of view, gathers more light and has a greater depth of focus (meaning that a greater thickness of specimens will appear to be in focus at any given time). At the other end of the spectrum is the 40× objective lens which has the greatest resolving and magnifying power but also has the narrowest field of view, gathers the least amount of light and has the shallowest depth of focus.

- **Care of light microscopes:** When moving microscopes, always carry them by securely holding both their base and arm. Avoid tipping microscopes or carrying them on their side to prevent ocular lenses and filters from falling out and breaking. Place microscopes on clean level surfaces and far from the edge of desks and tables. If your microscope becomes wet or soiled, use a paper towel and lab tissues to immediately clean and dry its surfaces. Never clean lenses with anything but the **lens paper** that will be provided and ask your teaching assistant for help in cleaning particularly dirty lenses. And, never force any of the mechanical controls on a microscope—when adjusted correctly they will all work freely.

STEP ONE

1. Work in pairs. Obtain a microscope from the microscope cabinet and place it securely on your desk. Plug its electrical cord into one of the electrical outlets on your laboratory bench.

2. Move the stage of your microscope to a low position with its coarse adjustment. Rotate the nosepiece of your microscope to position the 4× objective lens above the hole in the center of the stage.

3. Obtain a glass slide from the supply table in the front of the lab containing two or three strands of colored thread. Place the slide in the center of the stage of your microscope and secure it with the stage clips.

4. Rotate the light control knob (on the left side of the base of your microscope) to the "on" position and adjust it to medium intensity. Use the controls of the micrometer (immediately to the right and below the stage) to move the slide so that you can see light illuminating the threads on the slide. Use the coarse adjustment control of the microscope to move the stage and the slide closer to the objective lens (but don't bring them into contact with each other).

5. Begin looking through the eyepieces of the microscope. Rotate the ocular tubes (to which the eyepieces are attached) inward or outward so that one ocular is directly in line with each of the pupils of your eyes. When the ocular tubes are adjusted correctly you should see a single brightly illuminated circle of light. Adjust the light control, iris diaphragm and condenser position until the circle of light is uniformly illuminated from edge to edge.

6. While looking through the eyepieces, bring the threads into focus by gradually moving the stage *downward* using the coarse adjustment control. It may be necessary for you to use the micrometer to bring the threads into the field of view. Once the threads are in view use the fine adjustment control, light control, iris diaphragm and condenser to make the image of the threads you see as clear and sharp as possible.

7. After you and your laboratory partner have practiced focusing on and viewing the threads, carefully rotate the 10× objective into the viewing position. Whenever you rotate the lens of your microscope, watch to be certain that the objective lenses do not hit the slide or the stage. Because your microscope is **parfocal,** the focusing you have done with one objective lens should cause all the other lenses to be very close to being in focus as well when they are moved into position. Even still, you may find you need to make small adjustments with the fine adjustment control and micrometer when viewing with the 10× objective.

8. Repeat the previous point (7) of this experiment using the 40× objective lens. The light intensity you perceive with this lens will be very much lower and it will be necessary for you to make major adjustments to the light control, iris diaphragm and condenser.

9. Make a note of which of the three objective lenses you have used: (a) gives the best overview of the threads, (b) allows you to see the greatest detail of the substructure of each thread, (c) is most useful for finding threads to look at in the first place. Return the nosepiece to the position that allows viewing with the 4× objective lens. Remove the slide with colored threads from your microscope's stage and return it to the supply table.

STEP TWO

1. Obtain a clear plastic ruler calibrated in centimeters and millimeters from the supply table and place it on the center of your microscope's stage. Use the 10× objective lens and adjust the position and focus of the ruler so that its edge is visible across the center of the field of view.

2. Use the ruler to estimate to the nearest 0.2 millimeters (mm) the diameter of the field of view of your microscope and make a note of your observation.

3. The field diameters of two different lenses is inversely proportional to the magnification power of the two lenses. Consequently,

$$\text{field diameter}_H = \frac{\text{field diameter}_L \times \text{magnification}_L}{\text{magnification}_H} \tag{4.1}$$

    where "field diameter$_H$" and "field diameter$_L$" are the field diameters of the high and low powered objective lenses, respectively, and "magnification$_H$" and "magnification$_L$" are the magnification powers of the high and low powered objective lenses, respectively. Use the provided equation to determine the field diameter of the 40× and 4× objective lenses. Convert these estimates of field diameter to micrometers (μm are a much more common unit of measurement in microscopy) and report them in the results section of your laboratory report (1 mm = 1,000 μm).

4. Obtain a slide of a stained smear of human blood from the supply table and follow the procedures in Step One to view doughnut-shaped red blood cells. Red blood cells are unusual eukaryotic cells in that they are small and have no nuclei or other structures enclosed by internal membranes. Still, they have a relatively uniform size of approximately 6 μm and provide excellent practice in focusing on small objects with high magnification powers on your microscope.

5. When you have successfully viewed human red blood cells and confirmed that their size is approximately 6 μm, obtain a slide containing *Paramecium caudatum* cells from the supply table and follow the procedures in Step One to view representatives of this single-celled eukaryote. Knowing the field diameter for each of the objective lenses estimate the length of ten *Paramecium caudatum* cells to within 10 μm and record your estimates in Table 4.1.

Table 4.1 Microscopic size estimates of single eukaryotic cells.

Cell Examined	*Paramecium caudatum* Cells (length)	*Elodea* Cells (length)	*Elodea* Cells (width)
1			
2			
3			
4			
5			
6			
7			
8			
9			
10			
Average			

STEP THREE

1. Return the slides containing a stained human red blood cell smear and *Paramecium caudatum* cells to the supply table and obtain a clean microscope slide. Place a single *Elodea* leaf in a small drop of water on the slide. Use a dissecting needle or pencil point to place a clean cover slip at a 45° angle above the slide and in contact with an edge of the drop of water. Lower the cover slip slowly onto the slide while carefully avoiding trapping air bubbles underneath it.

    The preparation you have just made is known as a **wet mount.** The purpose of the cover slip is threefold: (a) it flattens the specimen you will be examining, (b) it prevents the preparation from drying out, and (c) it protects the objective lens from contact with specimens.

2. Place your wet mount on the stage of your microscope and view the *Elodea* leaf with each of the three objective lenses. Estimate the longest and shortest dimensions of the rectangular *Elodea* cells (their length and width, respectively) and record them in Table 4.1. The small, green objects within each cell are chloroplasts and the single round and dark bodies are nuclei. Include a completed version of Table 4.1 and a sketch of an *Elodea* cell in the results section of your laboratory report.

3. Discard your cover slip and *Elodea* leaf in the designated container on the supply table. Clean your slide with distilled water and gently flick it to remove most of the residual water. Place a drop of pond or aquarium water on your slide and use a cover slip to make a wet mount of it as you did in the first point (1) of this step of the experiment. Use your microscope to explore the diverse living things on your wet mount.

4. Make a sketch of at least three different kinds of organisms you find on your wet mount and include them in the results section of your laboratory report. Try to identify as many organisms as possible with the help of your teaching assistant and the other resources available in the laboratory.

5. Clean and return your slide to the blue tray at the end of your time in the laboratory for this experiment. Store your microscope in the cabinet from which you took it at the beginning of today's laboratory work.

IMPORTANT QUESTIONS TO CONSIDER

Careful consideration of the following questions related to this set of experiments will not only insure that you understand the most important concepts of this laboratory but will also help you prepare for the lecture portion of the course. By addressing these questions (and others that you decide are important) in the conclusions section of your laboratory report, you will also assure that you receive full credit for that portion of your report.

- HOW DO THE three objective lenses on the microscope you used differ and what are the advantages and disadvantages of each?

- HOW VARIABLE WERE the lengths of *Paramecium caudatum* cells relative to the lengths and widths of the cells you observed in *Elodea* leaves?

- WHAT SIZE RANGE did the single eukaryotic cells you examined span?

- WHAT SUBCELLULAR FEATURES did you recognize and identify in the organisms you examined?

- DID YOU FIND any prokaryotes in the pond or aquarium water you used?

- HOW WOULD PROKARYOTES and eukaryotes differ when viewed with a microscope?

- WHAT ORGANISMS WERE you able to identify in the pond or aquarium water you examined and how did they differ?

- TURN IN your observations from Step One, item 9, along with Table 4.1 and your sketches from Step Three, item 4.

EXPERIMENT 5:
ENZYMES AND FERMENTATION

Enzyme History: The term "enzyme" was first used by F.W. Kuhne in 1878 and was derived from a Greek phrase meaning "in yeast." By 1940 it became an official work in virtually every language and applied to all biological catalysts, not just those from yeast.

OVERVIEW

One of the most distinguishing features of living things is their ability to manipulate their internal environment by controlling the chemical reactions that occur within them. The primary means by which that control is achieved is through the use of biological catalysts called **enzymes.**

Among the most important of the reactions catalyzed by enzymes are those used to catabolize sugars and starches. The **fermentation** of glucose (glycolysis) to end products such as ethanol and lactic acid serves as the main energy-yielding catabolic route for most microbes in conditions where the supply of oxygen is limited. Like all enzymes, those involved in fermentation function best only within narrow pH and temperature ranges. Organisms go to great lengths to maintain internal environments that allow optimal activity of their enzymes.

In the course of this set of experiments, you will examine the activity of an enzyme involved in the break down of starch into its component sugars as well as the affect of fluctuations in temperature and pH on the enzyme's ability to act as a catalyst. In addition, you will directly assay the ability of living cells to benefit from the enzymes they possess as they extract usable energy from a set of energy-rich carbon molecules. Enzymes themselves are rarely visible even with the aid of the strongest microscopes. Still, in this set of experiments you will not only detect their presence but also examine the way they function and the roles they play in all living organisms.

INTRODUCTION

Catalysts increase the rate of chemical reactions by lowering the **activation energy** that must be put in to the molecule(s) to get the reaction started (Figure 5.1). Enzymes are biological catalysts, and each living cell has between hundreds and thousands of different enzymes that each specifically act upon different **substrates.** Each substrate is modified during the course of the reaction to form a specific product but the enzyme emerges unchanged and ready to act catalytically upon another substrate. Consequently, small amounts of enzymes can affect relatively enormous amounts of substrate.

After many years of work, J. B. Sumner reported the first isolation of an enzyme in 1926. His report that the urease he had purified from jack beans was a protein was greeted very skeptically by many of his colleagues. By 1935, however, the work of Sumner, as well as J. Northrop and W. M. Stanley, convinced all scientists that most enzymes were indeed proteins and earned these three pioneers in their field a Nobel prize in 1946.

The enzymes characterized in these initial studies and all since have three characteristics that distinguish them from all other natural and man-made catalysts on earth. First, they are the *most efficient catalysts known:* most cellular reactions take place at a rate over a million times faster than they would have in the absence of enzymes. Second, the majority of enzymes have an impressive *specificity of action:* Many enzymes act on only one substrate to yield only one product. Third, the actions of most enzymes are *regulated:* Cells can control their metabolic processes by changing individual enzymes from a state of low activity to one of high activity.

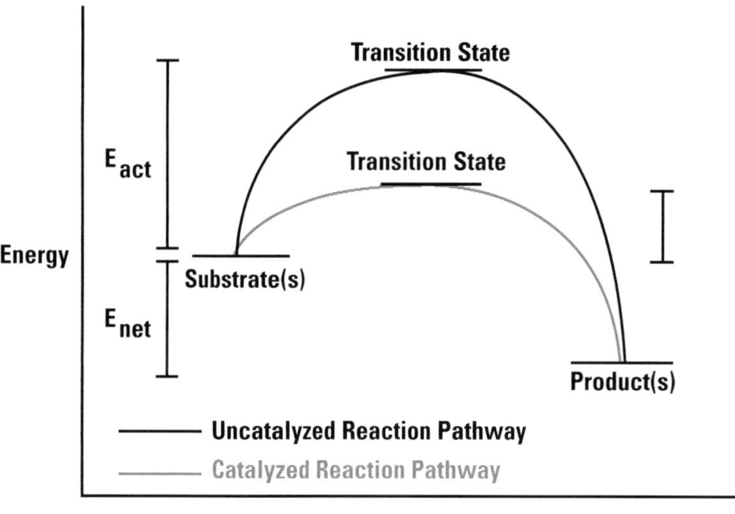

Figure 5.1 Energy profiles of unanalyzed and catalyzed reactions. The difference between the energy levels of the ground state of the reactants and the transition state is labeled as E_{act}. E_{net} corresponds to the difference between the substrates and products. Catalysts have dramatic effect on E_{act} but not on E_{net}.

The structure and function of enzymes are intimately related. Most enzymes are globular proteins whose convoluted folding brings specific amino acids into close proximity to form an **active site** at which substrates are specifically bound and converted to products. The specific interactions between the shapes of the active site of an enzyme and its substrate are often likened to those between a lock and key. By inducing very small changes in the shape of an enzyme, cells can have a dramatic effect upon its ability to bind substrates and generate products.

Enzyme shape is influenced by a large number of factors. Almost all antibiotics, insecticides, herbicides, poisons, and drugs that combat pain, inflammation, and cancers have their effects due to their ability to **inhibit** (interfere with) the function of specific enzymes by changing their shape.

Alterations in temperature and pH can dramatically effect the activity of enzymes in much the same way. Low temperatures often do not provide enough energy for even the lowered activation energy required by enzymatic catalysts to be achieved. At the same time, temperatures above 40°C cause many proteins to become **denatured** and lose their normal **tertiary** (three-dimensional) structures. And, recall that the R-groups of many of the amino acids from which proteins are built readily lose and gain hydrogen ions as pH rises and falls—the resulting changes in local charge typically have dramatic effects upon a protein's tertiary structure by forcing apart or pulling together regions on the surface of the enzyme.

The internal environment of cells is usually carefully regulated to maintain optimal conditions that allow most of the cell's enzymes to function at their maximum levels of activity. Still, as you would probably expect, not all substrates that cells encounter are catalyzed at the same rates—even when they are energy-rich carbon molecules. Many enzymes are regulated by cells in such a way that **preferred substrates** (typically those that can yield the most energy and require the fewest number of unusual enzymatic modifications) are catalyzed first while other substrates may never be used unless no others are available.

PROCEDURES

In the course of the first set of experiments in this laboratory, you will investigate the activity of amylase: an enzyme involved in the digestion of starch. Large amounts of amylase can be isolated from the saliva of humans as well as most other animals that regularly encounter starch in their diet. Starch molecules are actually polysaccharides composed of large numbers of glucose molecules linked together and are the primary molecule in which plants store the energy they derive from photosynthesis. Amylase cleaves the long chains of glucose in starch into two-glucose-long pieces (the disaccharide maltose). The complete digestion of starch by pancreatic and intestinal enzymes cannot occur if amylase does not perform this essential first step.

You will follow the conversion of salivary amylase's substrate (starch) into its product (maltose) at a range of pHs and temperatures by taking advantage of the fact that starch, but not maltose, turns a dark purple color when treated with potassium iodide (I_2KI).

STEP ONE

1. Work in pairs. Obtain five clean test tubes and a rack from your teaching assistant and label the reaction tubes "4," "5," "6," "7," and "9." Use re-pipettors to deliver 5 mL of five buffered solutions prepared by the prep-room personnel to the tube whose number corresponds to the solution's pH. These solutions, particularly those with the extreme pHs, can burn skin—handle them carefully and clean any spills with paper towels as quickly as possible.

2. Use a disposable pipette to transfer one drop of a 1% amylase solution to each of the five labeled reaction tubes. Mix the contents by gently swirling the reaction tubes.

3. Obtain a welled-plate and label it as shown in Figure 5.2. Use a disposable Pasteur pipette to transfer one or two drops of the yellow colored I_2KI solution into each well of the plate you have labeled.

    It is important that the time between the next point (4) and the one that follows it (5) in this experiment be as short as possible. Read those points and be certain you understand and are prepared for them before beginning the next point (4).

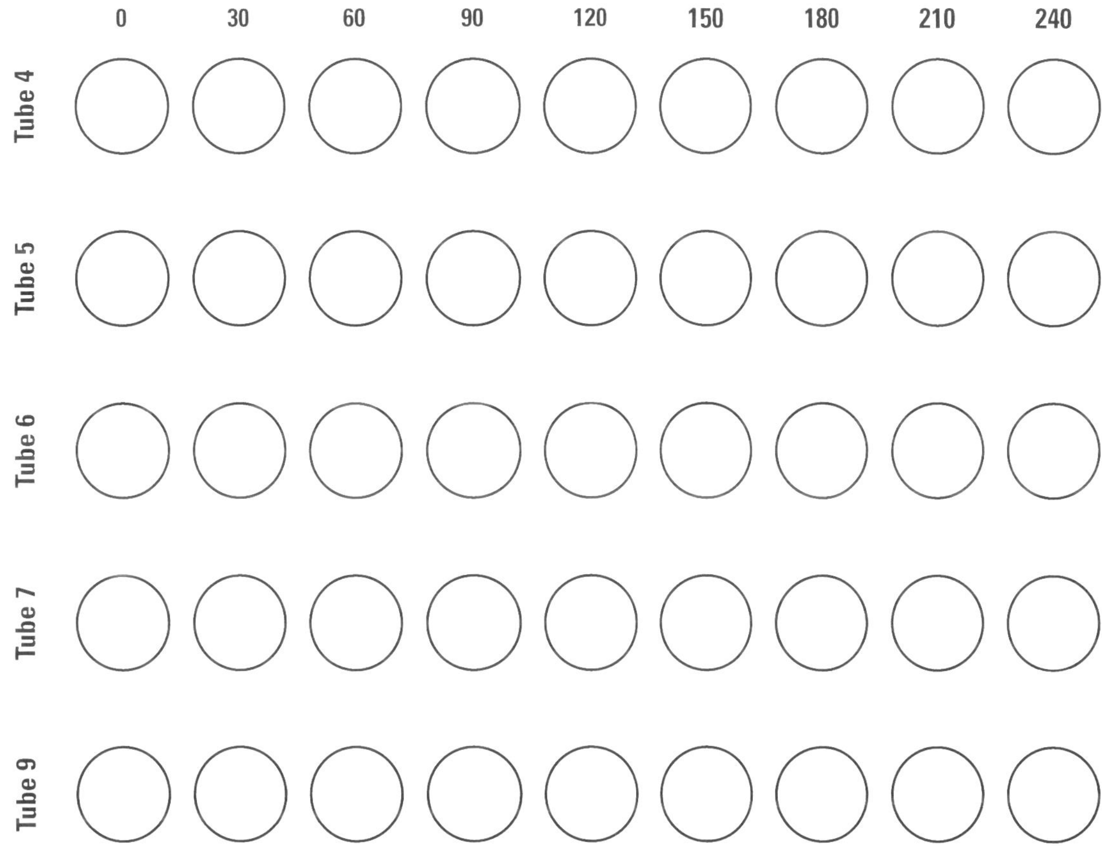

Figure 5.2 Representation of the welled plate containing potassium iodide. Rows are labeled with reaction tube numbers (reaction *p*Hs) and columns are labeled with time points (in seconds).

4. Using a clean pipette, transfer 2.5 mL of the 1% starch solution to a test tube. Gently swirl each tube to mix their contents. Combine the contents of the test tube containing the 2.5 mL of starch and the reaction tube labeled **4**. Make a note of the time (T_0) and mix the contents quickly by gentle swirling. Proceed as quickly as possible to the next point (5) of the experiment.

5. Use a disposable pipette to remove approximately three drops from reaction Tube 4. Add this drop to the drop of I_2KI in column "0," row "Tube 4" on your welled plate. Repeat this procedure every 30 seconds using a new spot of I_2KI on the welled plate until a very dark purplish-blue color is no longer produced (indicating that all the starch in the solution has been converted to maltose) or four minutes have elapsed.

6. Record the time at which starch was no longer present in the reaction tube in Table 5.1. Be sure to include this data in the results section of your laboratory report.

Table 5.1 Time of starch disappearance (in seconds).

pH of Reaction	Tube	Time (seconds)
	4	
	5	
	6	
	7	
	8	

7. Repeat the previous three points (4, 5, and 6) for each of the remaining four reaction tubes individually.

8. Rinse the welled plate and test tubes that you used after you have collected your data. After leaving the laboratory, use the data you recorded in Table 5.1 to construct a graph that plots the reaction pH *versus* the time needed for the starch to be converted to maltose. Be certain to include the graph in the results section of your laboratory report.

STEP TWO

1. Work in pairs. Clean the test tubes you used in the previous step by rinsing them with water and obtain three new tubes from your teaching assistant so that you have a total of eight clean test tubes. Divide these tubes into two sets and label each set "0," "22," "37," and "45." Use re-pipettors to deliver 5 mL of the 1% starch solution to one set of the four tubes and 250 μL of 1% amylase to each of the tubes in the other set.

2. Obtain a beaker large enough to hold two test tubes from your teaching assistant and fill the beaker with crushed ice.

3. Label a welled plate using Figure 5.3 as a guide. Use an eye-dropper to transfer one drop of the yellow colored I$_2$KI solution into each well of your labeled welled plate.

    Once again, it is important that the time between the next point (4) and the one that follows it (5) in this experiment be as short as possible. Read those points and be certain you understand and are prepared for them before beginning the next point (4).

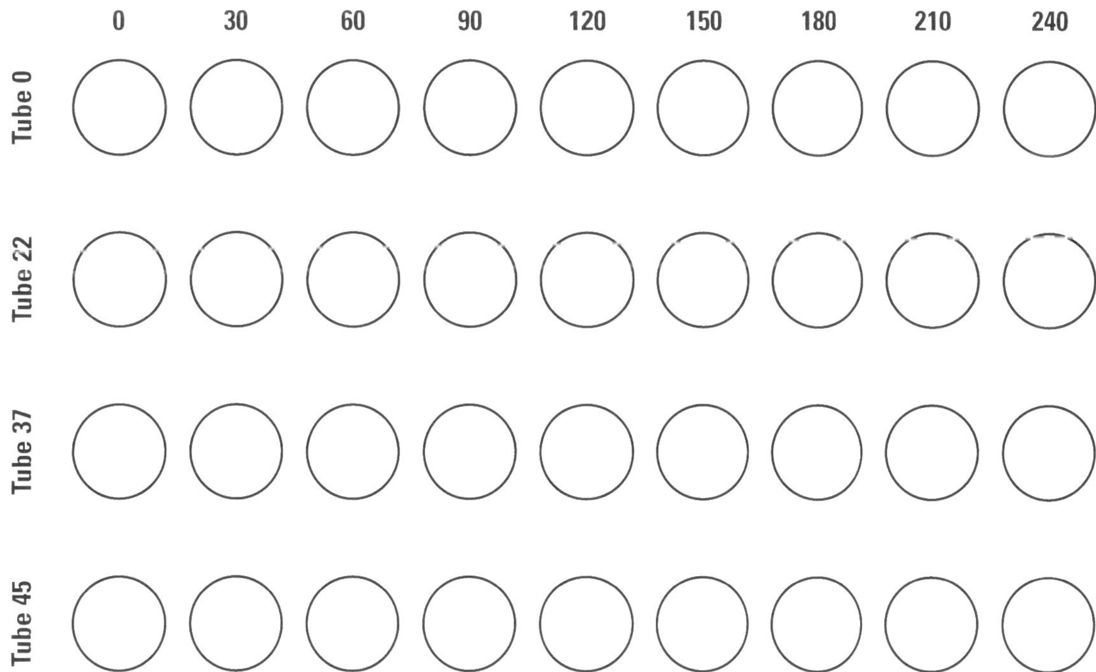

Figure 5.3 Representation of the welled plate containing potassium iodide. Rows are labeled with reaction tube numbers (reaction temperatures) and columns are labeled with time points (in seconds).

4. Place each of the tubes labeled "0" in your beaker containing crushed ice and let them come to that temperature for at least three minutes. Pour the contents of the "0" 5-mL starch tube into the "0" amylase tube and mix the solutions. Return the tubes containing the reaction to the ice as quickly as possible and proceed as quickly as possible to the next point (5) of the experiment.

5. Use a disposable Pasteur pipette to remove two drops from tube "0." Add this drop to the drop of I_2KI in column "0," row "Tube 0" on the welled plate. Repeat this procedure every 30 seconds (keeping the tube at the reaction temperature the entire time) using a new spot of I_2KI on the welled plate until a very dark purplish-blue color is no longer produced (indicating that all the starch in the solution has been converted to maltose) or four minutes have elapsed.

6. Record the time at which starch was no longer present in the reaction tube in Table 5.2. Be sure to include this data in the results section of your laboratory report.

7. Repeat the previous three points (4, 5, and 6) for each of the remaining three sets of reaction tubes substituting water baths with temperatures that correspond to their labels for each.

8. Wash the welled plate being careful to avoid contact with the solutions in the wells. Rinse and return the test tubes you have been using to your teaching assistant. After leaving the laboratory, use the data you recorded in Table 5.2 to construct a plot of reaction temperature *versus* the time needed for the starch to be converted to maltose. Be certain to include the plot in the results section of your laboratory report.

Table 5.2 Time of starch disappearance (in seconds).

Reaction Temperature (°C)	Time (seconds)
0	
22	
37	
45	

STEP THREE

Yeast *(Saccharomyces cerevisiae)* is a simple, unicellular eukaryote that can be classified as a facultative anaerobe because it can live in **aerobic** and **anaerobic** environments. In the absence of oxygen (anaerobic conditions),j yeast derives energy from carbohydrates by fermenting them and producing ethanol and carbon dioxide as waste-products. Since two molecules of carbon dioxide (a gas) and ATP are produced from every molecule of glucose (a solid) consumed in this way, it is possible to determine how much energy actively growing yeast obtains from three different carbohydrates and an artificial sweetener by monitoring the rate at which gas is produced.

A brief introduction to the molecular structure and role of each of the four compounds you will be presenting to actively growing yeast cultures should help you interpret the results of the following experiment:

- **Glucose** is a simple six-carbon sugar that is the primary molecule used for energy storage and transport in most organisms. While ATP can be likened to the one dollar bill of a cell's energy budget, glucose fills the role of a ten or twenty dollar bill and all cells have the enzymatic machinery in place to make rapid and efficient use of it.

- **Galactose** is another common six-carbon sugar but one whose molecular structure is generally converted into a form of glucose before cellular glycolytic enzymes extract the energy stored in it. The conversion of galactose to glucose-6-phosphate takes place only with the help of several enzymes and cells are not always prepared to utilize this equally good but less preferred energy source.

- **Sucrose** (or, table sugar) belongs to a class of carbohydrates known as disaccharides. Each sucrose molecule is actually two six-carbon monosaccharides (glucose and fructose) held together by a chemical bond known as a glycosidic linkage. Plants such as beets, sugarcane and maple trees, produce large amounts of sucrose and are able to readily hydrolyze it into glucose and fructose. Yeast also appreciate its sweet taste and recognize it as a valuable, relatively easily utilized carbon and energy source.

- **Nutrasweet**™ is not a carbohydrate at all but rather an unusual arrangement of three amino acids that was accidentally found to very effectively trick the chemoreceptors in mammalian tastebuds into behaving as if they were in the presence of energy-rich carbon molecules such as sucrose. While this compound may taste intensely sweet to us, very little useful energy can be extracted from this unusual tri-peptide.

1. Work in pairs. Obtain a ruler, five large test tubes, and five small test tubes from your teaching assistant and place them in a rack on your laboratory bench. Label the small test tubes 1 through 5.

2. Add 10 mL of boiled yeast suspension to the small tube labeled "1" and 10 mL of active yeast suspension to the remaining four small tubes.

3. Fill Tubes 1 and 2 up to 1 cm from the top with the 5% glucose solution provided by the prep-room personnel. Fill Tube 3 up to 1 cm from the top with a 5% galactose solution. Fill Tube 4 with a 5% sucrose solution. Fill Tube 5 with the 5% Nutrasweet™ solution.

4. Hold the small test tube labeled "1" upright. Invert a large test tube so that the bottom side is upward and lower it to cover the smaller test tube. Hold the test tubes together at their ends and invert the apparatus so that the small tube is now upside down inside the larger one.

5. Repeat the previous point (4) of this experiment for each of the remaining four small test tubes. If there is an air space at the top of any of the small tubes, measure the distance between its bottom and the top of the small test tube and record it as the "0" measurement for that tube in Table 5.3.

6. Measure the length of the gas column (the distance between the top of the small test tube and the bottom of the bubble) at the top of the small tube at 10 minute intervals for as long as possible (but not more than one hour) and record their heights in Table 5.3.

Table 5.3 Fermentation of compounds by yeast cultures.

Test Tube		Gas Column Height (measured at 10-minute intervals)						
Number	Contents	0	10	20	30	40	50	60
1	Boiled Yeast, Glucose							
2	Active Yeast, Glucose							
3	Active Yeast, Galactose							
4	Active Yeast, Sucrose							
5	Active Yeast, Nutrasweet™							

7. After the laboratory period is over, create a graph that includes plots of the height of the gas column *versus* time for each of the five samples you tested and be certain to include it and a completed version of Table 5.3 in the results section of your laboratory report.

IMPORTANT QUESTIONS TO CONSIDER

Careful consideration of the following questions related to this set of experiments will not only insure that you understand the most important concepts of this laboratory but will also help you prepare for the lecture portion of the course. By addressing these questions (and others that you decide are important) in the conclusions section of your laboratory report, you will also assure that you receive full credit for that portion of your report.

- WHAT IS THE optimal pH for salivary amylase activity and how does it compare with the pH at which the enzyme is normally found?

- THE pH INSIDE most animal's stomachs is usually between 3 and 4. How much dietary starch is likely to be converted to maltose in the stomach relative to the mouth and esophagus?

- WHAT IS THE optimal temperature for salivary amylase activity and how does it compare with the temperature at which the enzyme normally acts?

- WHAT VARIABLES OTHER than temperature and pH might alter the activity of salivary amylase?

- HOW DIFFERENT WERE the hydrogen ion concentrations of the buffers you used?

- WHICH OF THE four compounds you tested gave living yeast cells the most energy during the course of your experiment and why?

- HOW CONFIDENT ARE you in the results of your experiments and what factors may have led to errors?

EXPERIMENT 6:
ENZYME PURIFICATION AND ASSAY

"I think if people are passionate about something, it could be real estate or biochemistry, and that spark gets turned on in them, everyone's beautiful in that zone." Cindy Crawford

OVERVIEW

Living things are bound by the same laws of thermodynamics that govern chemical reactions in the world around us. A very wide range of chemical reactions occur spontaneously inside both living and non-living things all the time. The vast majority of those spontaneous reactions are **not** helpful to living things. A very important difference between living and non-living systems though is that living things make it much easier for certain chemical reactions to occur. These favored reactions are catalyzed by enzymes.

Enzymes are biological catalysts. All catalysts work by speeding up the rate at which a chemical reaction occurs. That increase in speed happens because they reduce a reaction's activation energy (the energy that substrates need to pass through a transition state before being converted to products). Enzymes do not change the overall amount of energy associated with a chemical reaction—spontaneous reactions always go from higher energy/order to lower energy/order.

Most enzymes are proteins that are made by the cells that use them. As you demonstrated in the last laboratory exercise, enzymes are affected by changes in temperature and pH. In this experiment you will use other features of a specific enzyme, wheat germ acid phosphatase, to separate it from all the other chemicals in an organism. Similar strategies are used to purify and study all commercially and medically important proteins like insulin, parathyroid hormone, and human growth hormone.

INTRODUCTION

The single best way to characterize the structure and function of a molecule is to start with a large amount of purified material. Hemoglobin, the protein that carries oxygen in our blood, was one of the first proteins to have the sequence of its amino acids determined largely because it was easy to obtain in large quantities in fairly pure form (almost 15% of the weight by mass of blood from slaughter houses is hemoglobin protein). But hemoglobin is an exception to the rule—the vast majority of proteins that organisms rely upon are at extremely low concentrations (less than 0.01% of a cell's mass) and are mixed with a wide variety of other proteins and macromolecules.

Biochemists are the scientists that develop an isolation strategy for proteins. Their work depends upon two things:

1. an assay that allows them to determine the presence and concentration of the protein of interest, and

2. the availability of precipitation procedures that maximize the amount of the chosen protein at the same time that it reduces the amount of other, unwanted molecules.

Assays are generally readily available for most enzymes and depend on the ability to detect a change (often in color) as substrates are converted to products as the reaction occurs. It is often convenient to talk about the amounts of an enzyme in terms of **enzyme units** (U)—typically the amount of enzyme that catalyzes the conversion of 1 micro mole (one millionth of a mole) of substrate to product in one minute at the temperature, pH, and substrate concentration that result in the greatest rate. In short, assays let a biochemist know if a purification step is increasing the concentration of an enzyme and if the enzyme is being damaged to the point that it can no longer function.

Precipitation procedures often begin with a series of trial and error experiments. There is no standard protein isolation technique that can be applied to all proteins and some large classes of proteins, like membrane-bound proteins, have so far been very resistant to commonly used approaches. The solubility of a protein in water is directly dependent on the hydrophobic and hydrophilic amino acids on its surface. Salts, such as ammonium sulfate or manganese chloride, can directly interact with charged amino acids on the surface of proteins and get in the way of water molecules that keep the protein dissolved. Proteins with relatively few hydrophilic amino acids on their surface can come out of solution in relatively low concentrations of salt as their hydrophobic patches begin to interact with those on other molecules. Heat treatment, methanol or ethanol precipitation, and dialysis with semi-permeable membranes are other commonly used protein purification strategies that work in much the same way.

The protein that you will be isolating and characterizing in this exercise is wheat germ acid phosphatase. The enzyme's formal name is phosphate-monoester phosphodydrolyase (acid optimum) and its systematic name is E.C. 3.1.3.2. It catalyzes the release of inorganic phosphate (P_i) from a wide variety of substrates like glucose-6-phosphate and ATP as shown in Figure 6.1 and was first purified in 1960 by two biochemists (B. K. Joyce and S. Grisolia) who used a strategy very similar to the one that you will be using. Joyce and Grisolia chose to use wheat germ because it is inexpensive, readily available, and contains large amounts of the enzyme to support the high metabolic rate of germinating wheat seeds.

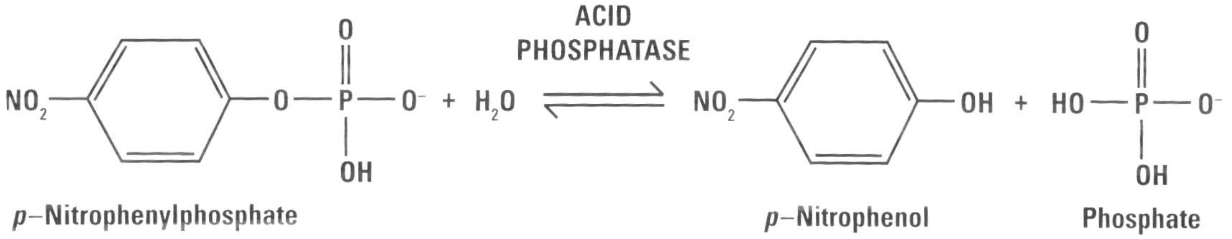

Figure 6.1 A chemical reaction catalyzed by the enzyme wheat germ acid phosphatase. The substrate for this reaction is p-nitrophenylphosphate and the products of the reaction are p-nitrophenol and inorganic phosphate (P_i). The substrate of the reaction has more Gibbs Free Energy than the products meaning that this is a spontaneous reaction. The reaction occurs about 1,700 times faster in the presence of the enzyme than its absence.

The purification begins with a water extraction in which wheat germ is soaked in cold, distilled water. A brief centrifugation gives rise to a supernatant that contains acid phosphatase as well as a wide variety of other water soluble molecules (e.g., nucleic acids, sugars, other proteins). Joyce and Grisolia, through trial and error, found that acid phosphatase is almost entirely soluble in a 35% ammonium sulfate solution but is not very soluble when ammonium sulfate concentrations are greater than 57%. That forms the basis of your second and third isolation steps.

There is an interesting lesson to be learned from the fact that biochemists are resigned to losing some of the protein they are purifying in every isolation step that they perform. As much as one third of the acid phosphatase remaining in your sample will be lost in both the second and third isolation steps. But, that is acceptable given that an even greater fraction of contaminating molecules are lost in each of those steps.

PROCEDURES

There are three inter-related parts to this laboratory exercise:

1. purification of acid phosphatase,

2. determination of the amount of protein present in several steps along the purification process, and

3. determination of the amount of acid phosphatase activity in several samples.

The purification strategy has four parts:

1. suspension of wheat germ in cold, distilled water,

2. manganese chloride precipitation,

3. 35% ammonium sulfate precipitation, and

4. 57% ammonium sulfate/heat precipitation.

Figure 6.2 shows a schematic of the overall strategy.

Step One (the enzyme purification) should be done as a team of four. Steps Two and Three should be done simultaneously after Step One is completed by two teams of two.

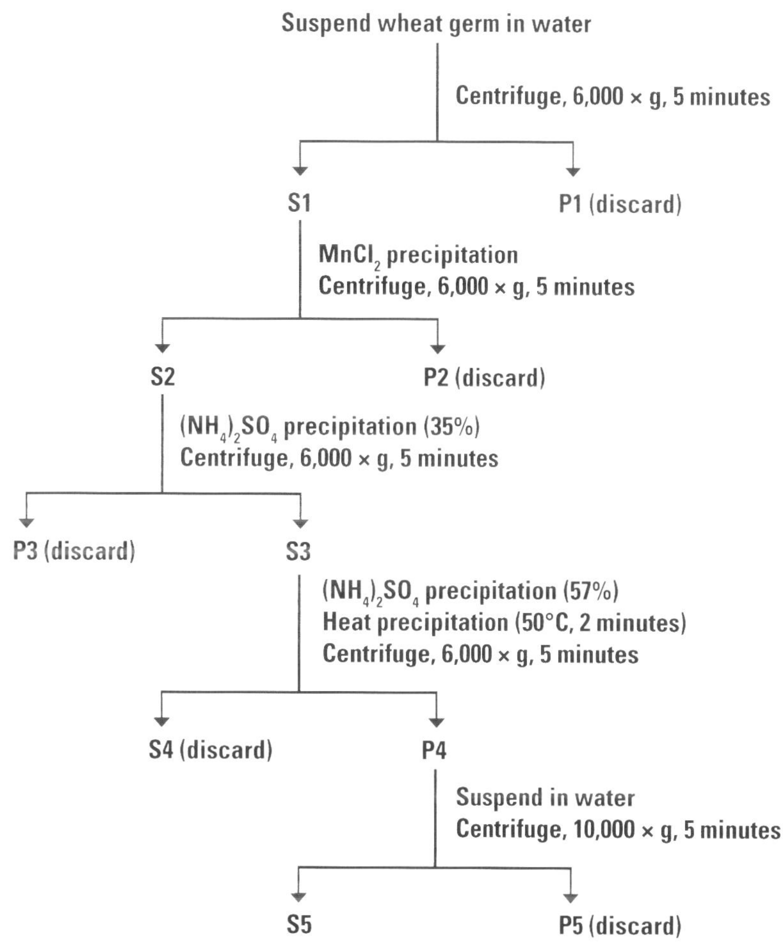

Figure 6.2 Overall strategy for wheat germ acid phosphatase purification. Either a pellet (P) or a supernatant (S) should be collected and used as the starting point for the next step in the purification.

STEP ONE

In this phase of the experiment, you will use a series of salt precipitations and a heat treatment to isolate and purify wheat germ acid phosphatase. Try to keep the flasks and cylinders that contain the enzyme cold as much as possible (except, of course, for the heat treatment) to preserve as much enzymatic activity as possible.

Wheat Germ Suspension

1. Work in teams of four. Label a single 500-µL microfuge tube "0" and add 300 µL of cold, distilled water to keep on ice and use as a blank later in your analyses. Observe a sample of raw wheat germ; and a sample of 5 g of raw wheat germ suspended in 10 mL of cold, distilled water (but not centrifuged). Obtain a 15-mL centrifuge tube that contains approximately 5 g of raw wheat germ that has been suspended in 10 mL of cold, distilled water for approximately 30 minutes before centrifuging at 6,000 × g for 10 minutes.

2. Carefully decant the supernatant (S1) into a chilled, graduated cylinder and record the total volume of liquid in Table 6.1. S1 contains the acid phosphatase enzyme and many other water-soluble molecules from wheat germ. Discard the pellet (P1) at the bottom of the centrifuge tube and clean the centrifuge tube for use in 5, below.

3. Transfer 300 µL of S1 to a cold 500-µL microfuge tube you have labeled "S1." Keep this tube on ice for protein quantitation and enzyme activity assays in Steps Two and Three of this experiment.

4. Transfer the remainder of S1 from the graduated cylinder to a 50-mL beaker in an ice bath. Gently stir while slowly adding 20 µL of 1 M $MnCl_2$ for every 1 mL of S1 in the beaker.

Manganese Chloride Precipitation

5. Transfer the $MnCl_2$-treated S1 to your cleaned (pellet-free) 15-mL centrifuge tube. Centrifuge at 6,000 × g for five minutes.

6. Carefully decant the supernatant (S2) from the centrifuge tube into a chilled, graduated cylinder. Record the total volume of S2 in Table 6.1. Clean the centrifuge tube (discarding the pellet, P2) for use in 9, below. S2 contains most of the acid phosphatase enzyme that was present in S1 while P2 contains many of the contaminating materials from the original extraction.

7. Transfer 300 µL of S2 to a cold 500-µL microfuge tube you have labeled "S2." Keep this tube on ice for protein quantitation and enzyme activity assays in Steps Two and Three of this experiment.

35% Ammonium Sulfate Precipitation

8. Transfer the remainder of S2 from the graduated cylinder to a 50-mL beaker in an ice bath. **Gently** stir while **very slowly** adding 0.5 mL of cold saturated ammonium sulfate for every 1 mL of S2 in the beaker. Pace yourself such that it takes about five minutes to add the total amount of ammonium sulfate so as to avoid locally higher concentrations within the beaker. Any foaming is an indication that proteins in S2 are being denatured and that the ammonium sulfate is being added too quickly or that stirring is too vigorous. At the completion of this process, your solution will be a 35% solution of ammonium sulfate.

9. Transfer the 35% ammonium sulfate solution to your cleaned (pellet-free) 15-mL centrifuge tube. Centrifuge at $6,000 \times g$ for five minutes.

10. Carefully decant the supernatant (S3) from the centrifuge tube into a chilled, graduated cylinder. Record the total volume of S3 in Figure 6.1. Clean the centrifuge tube (discarding the pellet, P3) for use in 14, below. S3 contains most of the acid phosphatase enzyme that was present in S2 while P3 contains many of the contaminating materials that still remained after manganese chloride precipitation.

11. Transfer 300 µL of S3 to a cold 500-µL microfuge tube you have labeled "S3." Keep this tube on ice for protein quantitation and enzyme activity assays in Steps Two and Three of this experiment.

57% Ammonium Sulfate Precipitation

The teaching assistant for your laboratory section may determine that there is insufficient time for you to complete points 12 through 17 of this process. If that happens, they will also tell you what sort of results you would have been likely to obtain.

12. Transfer the remainder of S3 from the graduated cylinder to a 50-mL beaker in an ice bath. **Gently** stir while **very slowly** adding 0.8 mL of cold saturated ammonium sulfate for every 1 mL of S3 in the beaker. As before, pace yourself such that it takes about five minutes to add the total amount of ammonium sulfate so as to avoid locally higher concentrations within the beaker. Any foaming is an indication that proteins in S3 are being denatured and that the ammonium sulfate is being added too quickly or that stirring is too vigorous. At the conclusion of this process, your solution will be a 57% solution of ammonium sulfate

13. Take the flask containing your 57% ammonium sulfate solution in a 60°C (hot) water bath for two minutes then immediately return the flask to an ice bath. Swirl the flask in the ice bath so as to reduce the solution's temperature to 5°C relatively quickly.

14. Transfer the heat treated, 57% ammonium sulfate solution to your cleaned (pellet-free) 15-mL centrifuge tube. Centrifuge at 6,000 × g for five minutes.

15. Carefully decant the supernatant (S4) from the centrifuge tube and discard it. Resuspend the pellet (P4) in cold, distilled water. Determine the amount of distilled water to use by dividing the volume you recorded for S3 in Table 6.1 by 3. Use a clean pipet tip to scrape and stir the pellet (P4). Centrifuge at 6,000 × g for five minutes.

16. Carefully decant the supernatant (S5) from the centrifuge tube into a chilled, graduated cylinder. Record the total volume of S5 in Table 6.1.

17. Transfer 300 µL of S5 to a cold 500-µL microfuge tube you have labeled "S5." Keep this tube on ice for protein quantitation and enzyme activity assays in Steps Two and Three of this experiment.

Table 6.1 Acid phosphatase isolation and protein quantitation.

Supernatant	Volume (mL)	Bradford Absorbance	Protein Concentration (ug/mL)	Total Protein (ug)
0				
S1				
S2				
S3				
S5				

STEP TWO

In this phase of the experiment, you will determine the concentration of protein in each of the supernatants (0, S1, S2, S3, and S5) that you have collected during the course of the protein isolation and purification you performed in Step One. You will determine protein concentrations using the "Bradford assay" which is based on an absorbance shift of the dye Coomassie Brilliant Blue G-250. Under acidic conditions the red form of the dye becomes its bluer form by binding to the proteins present in the solution being tested. The Bradford assay is less susceptible to interference by other chemicals than other protein assays but works best for a narrow range of protein concentrations (which often makes it necessary to perform dilutions of the sample being tested).

1. Work in pairs. Obtain five cuvettes and label them "0," "S1," "S2," "S3," and "S5." Add 10 μL from each of the corresponding tubes you obtained in Step One of this experiment plus 90 μL of distilled water. Recall going forward that this is a dilution.

2. Add 1500 μL of the Bradford dye reagent to each of your five cuvettes. Mix the contents of each tube and let rest at room temperature for five minutes. (Be careful—Bradford dye reagent will stain the protein in your hands and in your clothes.)

3. Use the blank to set the 0 measurement at 595 nm, then measure the absorbance of each of your experimental cuvettes at 595 nm. Record your results in Table 6.1.

4. Make a standard curve for the Bradford protein assay using data provided by the teaching assistant for your laboratory section. Use the standard curve to determine the concentration (μg/mL) in each of your five samples from Step One. If the absorption for one or more of your samples is greater than the highest value in the standard curve, then you will need to dilute those samples ten-fold (take 10 μL of your sample and add 90 μL of distilled water) and repeat points 1, 2, and 3 for those samples.

5. Calculate (multiply the concentration in ug/mL by the total volume of liquid recorded in Table 6.1; **remember to also multiply by 10 any sample that you needed to dilute)** the total amount of protein in each of the five samples for which you used the Bradford protein assay. Include Table 6.1 in the materials you turn in for this laboratory exercise.

STEP THREE

In this phase of the experiment, you will determine the amount of acid phosphatase activity in each of the supernatants (0, S1, S2, S3, and S5) that you have collected during the course of the protein isolation and purification you performed in Step One.

1. Work in pairs. Obtain five 1.5-mL microfuge tubes and label them "0," "S1," "S2," "S3," and "S5." Add 10 µL from each of the corresponding tubes you obtained in Step One of this experiment plus 90 µL of distilled water. Recall going forward that this is a dilution.

2. Add 480 µL of acid phosphatase assay reagent (containing sodium acetate, magnesium chloride, and p-nitrophenyl phosphate) to each of your five 1.5-mL microfuge tubes. Mix the contents of each tube and place them in a 37°C water bath for two minutes.

3. Add 10 µL from each of the five samples you set aside in Step One to the corresponding 1.5-mL microfuge tube you have incubated at 37°C water bath for two minutes. Mix each tube and return to the warm water bath for exactly five minutes. After five minutes have passed add 250 µL of 0.5 M KOH to each reaction tube. Mix the contents of each tube to stop any catalysis by wheat germ acid phosphatase.

4. Centrifuge each tube for two minutes. Transfer the supernatant (being careful to not disturb the precipitates) to disposable cuvettes and measure their absorbance at 405 nm. Record your results in Table 6.2.

5. Make a standard curve for acid phosphatase activity using data provided by the teaching assistant for your laboratory section. Use the standard curve to determine the enzyme activity (U/mL) in each of your five samples from Step One. If the absorption for one or more of your samples is greater than the highest value in the standard curve, then you will need to dilute those samples ten-fold (take 10 µL of your sample and add 90 µL of distilled water) and repeat points 1, 2, and 3 for those samples.

6. Calculate (multiply the enzyme activity in U/mL by the total volume of liquid recorded in Table 6.2; **remember to also multiply by 10 any sample that you needed to dilute)** and record in Table 6.2 the total amount of enzyme activity in each of the five samples for which you assayed acid phosphatase activity. Include Table 6.2 with the materials you turn in for this laboratory exercise.

Table 6.2 Acid phosphatase activity and purification.

Supernatant	Volume (mL)	Assay Absorbance	Enzyme Activity (U/mL)	Total Enzyme (U)
0				
S1				
S2				
S3				
S5				

A Note on Standard Curves

Steps Two and Three of this experiment both ask you to make and work with "standard curves." Standard curves are sometimes known as "calibration curves." Samples with known properties (here known amounts of amylase or known amounts of protein) are measured and graphed. The graphs then allow the amounts of amylase and protein in the samples you collected during the course of the experiment to be determined.

For example, for the Bradford assay you recorded an absorbance value or each of five different concentrations of protein. Make a graph that plots the known concentrations of protein on the X (horizontal) axis and the absorbance value on the Y (vertical) axis. Use this standard curve to then determine how much protein is in each of the samples you collected by seeing what concentration corresponds to the absorbance value you obtained for those samples.

IMPORTANT QUESTIONS TO CONSIDER

Careful consideration of the following questions related to this set of experiments will not only insure that you understand the most important concepts of this laboratory but will also help you prepare for the lecture portion of the course. By addressing these questions (and others that you decide are important) in the conclusions section of your laboratory report, you will also assure that you receive full credit for that portion of your report.

- HOW DID THE total amount of protein change as you proceeded through each step of this experiment? What fraction of the original amount of protein was present at the last step relative to when you began?

- HOW DID THE total amount of wheat germ acid phosphatase activity change as you proceeded through each step of this experiment? What fraction of the original amount of acid phosphatase activity was present at the last step relative to when you began?

- HOW MUCH DID you purify wheat germ acid phosphatase? (Divide the amount of enzyme activity per mg of protein at the end of the experiment by the amount of enzyme activity per mg of protein at the start of the experiment.)

- WHAT STEPS COULD you take to further purify wheat germ acid phosphatase?

- WHICH OF THE purification steps was the most effective?

- WHY WOULD IT not be helpful to simply repeat the most effective purification step several times?

- WHY WAS IT important for you to keep your samples cold until the enzyme activity assay steps?

- ACID PHOSPHATASES ARE produced by many human tissues such as the liver, kidney, spleen and prostate gland. The prostate gland enzyme, but not that of other tissues, is inhibited by tartrate ions. Prostate gland acid phosphatase levels in blood (determined with the same assay that you used) can be an important indicator of prostate cancer. How could this information be used to develop a human blood test for prostate cancer?

EXPERIMENT 7: PHOTOSYNTHESIS

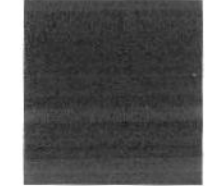

"I could always improvise. Some of my teachers remember me standing in front of the class with a flower on my head, talking about photosynthesis. I'd stop and say, 'Is this working for any of you?' The kids were like 'What is he doing?'" Bill Crystal

OVERVIEW

All living things on earth can be placed in to one of two categories: *autotroph* or *heterotroph*. Heterotrophs such as human beings drive their metabolic processes by harvesting the energy that other living things have stored in hydrocarbons. Autotrophs are capable of harvesting energy from non-living systems and storing it in hydrocarbons that they (or heterotrophs that eat them) can later use. Most autotrophs are *photoautotrophs* meaning that they harvest energy from light though *chemoautotrophs* (such as sulfo-bacteria that live around deep sea vents and get their energy from compounds like H_2S) are also worth mentioning.

Photoautotrophs capture the sun's energy through the process of photosynthesis—the production of organic material (hydrocarbons) and oxygen from carbon dioxide and water in the presence of light. Light excites electrons which are then passed through a membrane bound set of cytochromes where their energy is used to make energy-rich ATP and NADH molecules. Those molecules are then used to drive a process known as the Calvin Cycle where one-carbon molecules (carbon dioxide) are converted into three-carbon molecules that are eventually made into six-carbon molecules (glucose).

In the laboratory exercises that follow, you will be determining the wavelength of visible light that plants are best at converting into the chemical energy stored in the covalent bonds of sugars. You will also obtain evidence that supports the idea that chloroplasts are the site within cells that are responsible for the process of photosynthesis and the generation of energy-rich starch molecules.

INTRODUCTION

The simplest representation of the photosynthetic chemical equation is:

$$CO_2 + H_2O \rightarrow CH_2O + O_2. \tag{7.1}$$

Six repetitions would produce one six-carbon (C_6) glucose molecule.

One of the earliest and key insights into photosynthesis is that the *oxygen* that it produces comes from *water* and not from carbon dioxide. Interestingly, sulfobacteria use H_2S in the same way and produce sulfur gas (S_2) as a by-product much as plants produce oxygen gas (O_2). The reason this observation was so important is because the most important transfer of energy in photosynthesis occurs during the movement of hydrogens (and their electrons) from water (where their bonds have little energy associated with them) to hydrocarbons (where their bonds have much more energy associated with them).

Photosynthesis is therefore a redox process where the electrons associated with the hydrogens are forced to go from low potential energy to high.

To understand the energy transfers in photosynthesis though it is necessary to appreciate the photon packets in which it first comes to cells, the energy that comes from the sun is a major by-product of the fusion of hydrogen to form helium much like what occurs within a hydrogen bomb. A very small fraction of the energy that comes from those fusions falls to earth (it actually takes almost seven minutes for it to arrive) and only a small portion of that falls on the leaves of plants and is captured by photosynthesis.

Light is a form of energy known as electromagnetic energy. The energy is conveyed as the motion of an electromagnetic field in a wave. The closer together the crests of the wave are, the more energy is present within it. There is a very wide gamut of electromagnetic waves ranging from gamma waves (with wavelengths of 10^{-5} nm) to X-rays (1 nm) to visible light (380 nm to 750 nm; purple to yellow to red) to infrared (1 mm) to microwaves (1 m) to radio waves (kms). The shorter the wavelength, the more energy inherent in the photon of the wave.

The white light from the sun is actually the complete spectrum of wavelengths of visible light ranging from roughly 400 nm to 750 nm. Chlorophyll works best at absorbing light in two parts of the spectrum: red (about 680 nm) and purple (about 400 to 500 nm) as shown in the absorption spectra of chlorophyll in Figure 7.1. The result is chlorophyll's characteristic green color—green is the only light that is **not** readily absorbed when sunlight passes through a chloroplast (as a consequence, that is all that is left for us to see when the photons bounce off a leaf and hit our eyes).

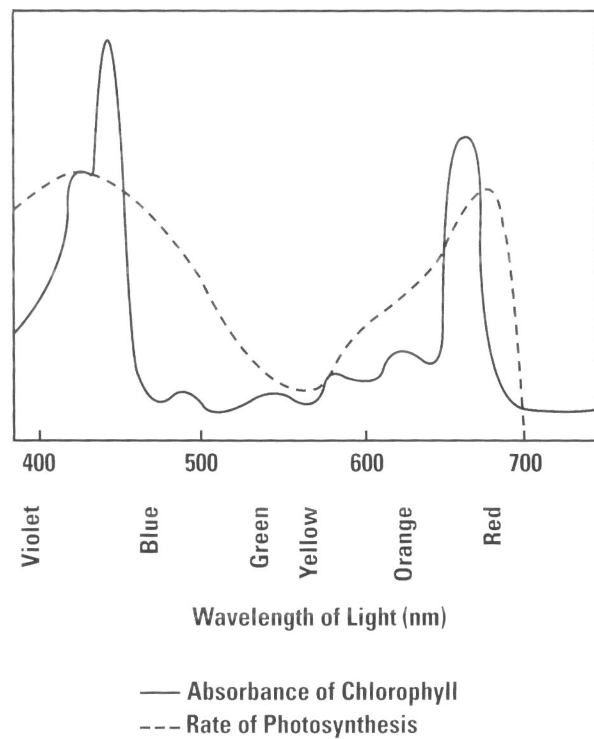

Figure 7.1 Absorption spectrum for photosynthesis and rates of photosynthesis for different wavelengths of visible light. Note that chlorophyll absorbs purple and red light best and barely absorbs any yellow and green light. Shorter wavelengths have photons with more energy than longer wavelengths.

STEP ONE

Early in the laboratory, prep-room personnel will have set up three strong lights of equal intensity on a laboratory bench. A large beaker containing water that is nearly ice cold will have been placed immediately in front of each light to act as a heat sink. A healthy green sprig of *Elodea* with an end freshly cut at a diagonal will have been inserted stem-end up in a large test tube at distances of 25 cm, 50 cm, and 75 cm from these heat sinks. In addition to the *Elodea,* each test tube will be filled with a 0.5% sodium bicarbonate ($NaHCO_3$) solution to act as a pure carbon source for the plants.

1. Observe the tube containing a sprig of *Elodea* that is furthest from the heat sink. If fewer than two bubbles of gas (freshly made oxygen gas, a by-product of photosynthesis as seen in Equation (7.1)) per minute appear at the end of the cut stem, notify your teaching assistant or the prep room personnel who will add a fresh sodium bicarbonate solution to all of the test tubes. If two or more gas bubbles per minute are seen, however, count the number that appear on the end of the stem over a three minute period and record that number in Table 7.1.

Table 7.1 Dependence of photosynthesis and oxygen production on light.

	Distance of *Elodea* Sprig from Heat Sink		
Trial	25 cm	50 cm	75 cm
1			
2			
3			
Average			

2. Before the sodium bicarbonate levels in each of the three test tubes has had an opportunity to become too different from each other, repeat the previous portion (1) of the experiment for the test tubes that are closer to the heat sink.

3. Repeat the first two points (1 and 2, above) of this experiment and record your observations as Trial 2 in Table 7.1. Perform as many additional complete sets of trials as time permits. Compute the average values for the trials at each of the three distances and include a completed version of Table 7.1 in the results section of your laboratory report.

STEP TWO

Just as for Step One, prep room personnel will have set up three strong lights of equal intensity but now of different colors (blue, green, and red) on a laboratory bench. A large beaker containing water that is nearly ice cold will have been placed immediately in front of each light to act as a heat sink. A healthy green sprig of *Elodea* with an end freshly cut at a diagonal will have been inserted stem-end up in a large test tube at distances of 25 cm from these heat sinks. As before, each test tube will be filled with a 0.5% sodium bicarbonate ($NaHCO_3$) solution to act as a pure carbon source for the plants.

1. Observe the tube containing a sprig of *Elodea* that is exposed to the blue light. If fewer than two bubbles of gas (freshly made oxygen gas, a by-product of photosynthesis as seen in Equation (7.1)) per minute appear at the end of the cut stem, notify your teaching assistant or the prep room personnel who will add a fresh sodium bicarbonate solution to all of the test tubes. If two or more of gas bubbles per minute are seen, however, count the number that appear on the end of the stem over a three-minute period and record that number in Table 7.2.

Table 7.2 Photosynthesis with different wavelengths of visible light.

Trial	Color of Light to Which *Elodea* Sprig Is Exposed		
	Blue	Green	Red
1			
2			
3			
Average			

2. Before the sodium bicarbonate levels in each of the three test tubes has had an opportunity to become too different from each other, repeat the previous portion (1) of the experiment for the test tubes that are exposed to the green and red lights (there may not be bubbles of oxygen gas coming from the *Elodea* in some of the tubes in which case you should record a value of "0" in Table 7.2).

3. Repeat the first two points (1 and 2, above) of this experiment and record your observations as Trial 2 in Table 7.2. Perform as many additional complete sets of trials as time permits. Compute the average values for the trials at each of the three distances and include a completed version of Table 7.2 in the results section of your laboratory report.

STEP THREE

The appearance of oxygen bubbles from the cut sprig of *Elodea* is a direct indication that photosynthesis has occurred. The occurrence of starch in the leaves of a plant is another reliable indication that photosynthesis has taken place. As you learned in "Experiment 2: Concentration, pH, and Buffers," potassium iodide (I_2KI) can be used to detect the presence of starch. Some plants have variegated leaves that are the same in all ways except that some regions contain chlorophyll and some do not. You will use such leaves to determine if regions with chlorophyll generate starch and regions without chlorophyll do not.

1. Obtain a variegated leaf from a coleus or croton plant provided for your laboratory section by a member of the prep lab staff.

2. Make a sketch of the variegation pattern on the leaf you have obtained in the space provided below in the (A) panel of Figure 7.2.

3. Use laboratory tongs to take your leaf to the laboratory's fume hood. Have your teaching assistant or a member of the prep lab staff help you to decolorize your leaf by immersing it for approximately one minute in a beaker containing gently boiling 90% ethanol. (90% ethanol is flammable—be careful.) The decolorization is complete when the leaf's cholorophyll has been dissolved and the leaf is translucent and a uniform shade of light yellow.

4. Place your decolorized leaf in a finger bowl (available for you at the laboratory's fume hood) and return to your work station.

5. Add 5 mL of potassium iodide (I_2KI) solution to your finger bowl that contains your decolorized leaf (remember, potassium iodide can stain your skin and your clothing). Make a sketch in the (B) panel of Figure 7.2 of the staining patterns that you see.

6. Compare panels (A) and (B) in Figure 7.2. Is there a correspondence between portions of your leaf that contained chlorophyll and portions that were stained by potassium iodide (I_2KI)?

Figure 7.2 Sketches of a variegated and of an iodide stained leaf. Panel **(A)** on the left shows the variegation pattern seen on the leaf before it was decolorized and stained, panel **(B)** shows the staining pattern observed on the leaf after decolorization and staining with potassium iodide.

IMPORTANT QUESTIONS TO CONSIDER

Careful consideration of the following questions related to this set of experiments will not only insure that you understand the most important concepts of this laboratory but will also help you prepare for the lecture portion of the course. By addressing these questions (and others that you decide are important) in the conclusions section of your laboratory report, you will also assure that you receive full credit for that portion of your report.

- WAS THE LEVEL of oxygen production you observed in Step One of the experiment dependent on the light intensity reaching the plants?

- WAS THE LEVEL of oxygen production you observed in Step Two of the experiment dependent on the wavelength of the light reaching the plants?

- HOW DID THE results of the experiment with white light at 25 cm compare to those for blue, green, and red light at the same distance? Is this what you would expect given the composition of white light?

- WHAT RESULTS WOULD you have expected in Step Two of this experiment if light bulbs that emitted *only* yellow light were used? Violet light?

- WAS THERE A correspondence between the portions of your leaf that contained chlorophyll and those that contained starch (as determined by a color change upon exposure to potassium iodide)?

- WHY MIGHT THERE not be starch in portions of leaves that do not have chlorophyll?

EXPERIMENT 8:
METABOLIC RATES AND THE CARBON CYCLE

OVERVIEW

In the previous experiment you observed first-hand how photosynthesis allows plants to convert the energy associated with visible light into energy-rich complex carbohydrates. Cells use that captured energy to drive a very wide variety of chemical reactions that would otherwise be energetically unfavorable. Reactions like these, and all the others that take place within a living organism taken together, are known as metabolism.

As you might expect from your study of the effect of temperature on enzyme activity in "Experiment 3: Osmosis and Diffusion," the activity and metabolism of animals whose body temperature is the same as their environment's is radically affected by changes in environmental temperature. Perhaps more of a surprise though is the fact that, ounce for ounce, there is also a very strong inverse relationship between an organism's overall metabolic rate and its size.

By monitoring the production of a metabolic end-product (carbon dioxide) common to all animals, you will confirm these relationships while you determine the relative metabolic rates of several organisms in this laboratory's experiments. In addition, you will examine the manner in which catabolism and anabolism are intimately associated in nature by observing the cycling of carbon molecules through plant and animal systems.

INTRODUCTION

Metabolism is the sum total of all the chemical reactions that occur in a living organism. It is a *vast* subject. The number of reactions involved alone is impressive: between several hundred and several thousand depending on the complexity of the organism being considered. Although each individual reaction has its own importance, the viability of the organism as a whole requires the integration of all the individual reactions into an intricate, tightly controlled maze of reactions and pathways.

Despite the complexity and billions of years of fine-tuning by natural selection, metabolism is notably *in*efficient. When living things extract energy from carbohydrates and fats, only about 40% of the total energy available in those molecules is captured in a useful form (usually ATP). More than half of the energy present in all the food that organisms metabolize is simply lost as heat.

Most organisms derive no appreciable benefit from that heat energy because it is quickly lost to their environment (particularly if they live in water). For these organisms (**poikilotherms**), the environment itself dictates their body temperatures. As a result, the metabolic and overall activity of these cold-blooded organisms is directly correlated with the temperature of their environments—as temperature increases (within narrow limits) they catabolize more carbohydrates and become more active and as it falls they catabolize less and become lethargic.

A few organisms such as mammals have evolved mechanisms that retard the loss of metabolic heat energy to the environment. These organisms (**homeotherms**) are capable of maintaining relatively constant body temperatures that are generally higher than that of their environments. As a consequence, these organisms enjoy the benefits of higher metabolic rates (constantly high activity levels and the ability to exploit varied habitats) as well as pay the price of needing to intake greater quantities of energy-rich molecules with which they maintain their accelerated metabolic processes.

In both homeothermic and poikilothermic organisms and in plants, normal metabolic rates are inversely related to body size (Figure 8.1). The necessity for this is easily understood in the case of homeotherms—smaller animals have a greater surface-to-volume ratio and therefore lose heat to their environment more quickly. To successfully maintain a constant high body temperature despite the rapid loss of heat across its body's surface, a small animal must constantly oxidize food at a fast rate. An extreme example is found in shrews whose tiny size (about 4 grams) necessitates their consuming nearly their own body weight in food each day. Small animals like shrews can actually starve to death after only a few hours without food.

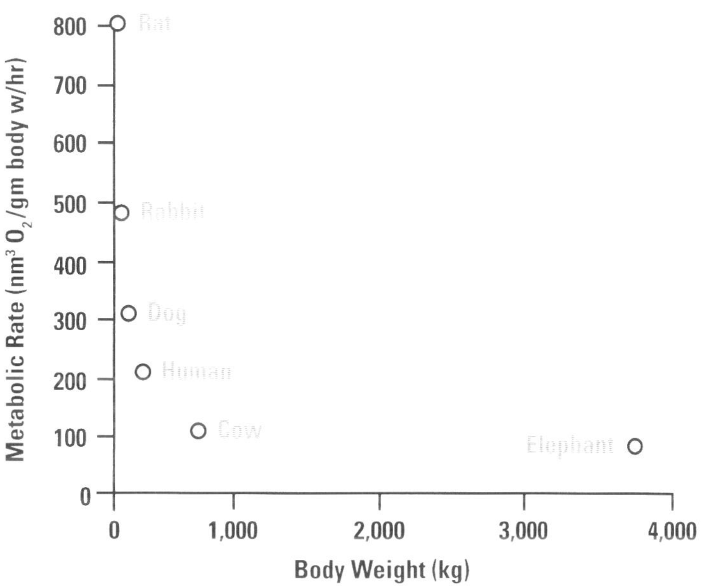

Figure 8.1 The relative metabolic rates of mammals are inversely related to the log of their body weights.

The inverse relationship between body size and metabolic rate in poikilothermic organisms is not as easy to explain. Because the surface-to-volume ratio of an organism becomes smaller as its size increases, large size in these organisms might be expected to result in reduced heat loss to the environment. That, in turn, should automatically result in a speeding up of their metabolic activity. However, increasing size usually involves a disproportionate increase in skeletal and supportive tissues as well. While these tissues contribute significantly to the mass of an organism, they are not metabolically active and reduce the apparent overall metabolic rate for large poikilotherms.

Almost all living things on earth use molecular oxygen in the process of **cellular respiration** to extract energy from end products of glycolysis (for example, the citric acid cycle as it occurs within eukaryotic mitochondria). **Heterotrophs** are organisms that are incapable of manufacturing organic compounds from raw materials such as carbon dioxide. Heterotrophs derive all their energy from the oxidation of organic nutrients produced by autotrophs such as plants. Most **autotrophs** use photosynthesis to build complex organic compounds from smaller, energy-poor molecules while they, like heterotrophs, also use oxidation to extract energy from the organic compounds they synthesize.

The metabolic rate of heterotrophs is relatively easy to determine simply by monitoring the rate at which they produce carbon dioxide (CO_2), the ultimate end product of cellular respiration (Figure 8.2). In contrast, the metabolic rate of autotrophs is typically more difficult to measure because carbon dioxide is also the primary small, energy-poor molecule used to assemble complex organic compounds in photosynthesis. As a result, in autotrophs the same simple molecule is simultaneously produced as a waste product and consumed as an essential nutrient.

$$C_6H_{12}O_6 + 6\ O_2 \leftrightarrow 6\ CO_2 + 6\ H_2O + \text{energy}$$

| Glucose | Oxygen | Carbon Dioxide | Water |

Figure 8.2 The overall process of cellular respiration as well as carbon fixation. Reactions proceeding from left to right constitute cellular respiration while reactions from the right to the left are those of photosynthesis (carbon fixation). The "energy" in cellular respiration is generally in the form of the high-energy chemical bonds of ATP and heat while in photosynthesis it is the electromagnetic energy of light.

Even though it confounds simple experiments designed to determine metabolic rates, this phenomenon is an excellent illustration of the **carbon cycle**—most biologically created carbon compounds are eventually returned to carbon dioxide, the molecule from which they were made. In nature complete passage through this cycle may take only minutes (if the compounds are oxidized by the plants that made them), days (if they are utilized by animals that eat the plants), or at most a few years (if they are ultimately utilized by animals several steps removed in the food chain from the original autotroph).

Of course, a small portion of carbon molecules escape this rapid turnaround and are captured by geological processes that convert them to fossil fuels such as coal, oil, and natural gas. Human beings have greatly augmented the return of this otherwise trapped carbon to the active carbon cycle by the burning of fossil fuels. Our activities have resulted in a 36% increase in atmospheric carbon dioxide levels since 1750. These increases in atmospheric carbon dioxide levels have been the subject of political debates over the past forty years. But scientists generally agree that the increase in atmospheric carbon has resulted in significant increases in the average temperature at the surface of our planet. They further agree that that trend is likely to continue in a way that will result in significant increases in sea levels and the patterns of precipitation and also have adverse effects on the distribution and number of species on Earth.

PROCEDURES

Carbon dioxide (CO_2) quickly interacts with water (H_2O) to become carbonic acid (H_2CO_3) in aqueous solutions. Carbonic acid is almost completely ionized ($HCO_3^- + H^+$) at neutral pHs. As a result, the input of carbon dioxide into an aqueous solution causes an increase in the concentration of hydrogen ions (H^+) and a consequent decrease in pH.

In this experiment you will use titration to measure the amount of carbon dioxide produced in a fixed period of time by a variety of organisms. By slowly adding a strong base, sodium hydroxide (NaOH), to water in which an organism has carried out its metabolic processes you will counter the pH-altering effects of the carbon dioxide it has released. When a pH indicator (phenolphthalein, see "Experiment 1: Dilutions and Standard Curves") that is also present in this water changes from colorless to bright pink (the **endpoint** of the titration) you will be able to determine how much carbon dioxide was produced by your experimental organism from the amount of sodium hydroxide used. Simply put, the more sodium hydroxide needed to raise your solution's pH and bring about the color change of phenolphthalein, the more carbon dioxide your organism produced.

STEP ONE

1. Work in groups of four. Obtain seven clean 500-mL flasks from your teaching assistant and label them "1," "2," "3," "4," "5," "6," and "7." Use a graduated cylinder to carefully add 200 mL of distilled water from the large carboy on the equipment table at the front of the laboratory to each of the seven flasks.

2. Weigh the first three flasks containing water using the scales provided on the equipment table and record their weights in the second column of Table 8.1.

3. Place a large snail from the aquarium on the equipment table in flask "1." Use a fishnet to obtain three crayfish from the aquariums on the equipment table and transfer one into each of the flasks labeled "2," "3," and "4" being careful to avoid traumatizing the crayfish or losing any of the water from your flasks.

4. Re-weigh each of the first three flasks that now contain water and your crayfish using the scales provided on the equipment table. Record the new weights of the flasks in the third column of Table 8.1. Ask your teaching assistant for help if any of the flasks now weigh less than or the same as they did in their first weighing (you've probably lost some water and will need to set that flask up again). Subtract the second weight from the first weight to determine the mass of your crayfish and record that weight in the fourth column of Table 8.1.

Table 8.1 Raw data for metabolic rates of experimental organism.

Flask	Contents	Initial Weight (g)	Weight with Crayfish (g)	Crayfish Weight (g)	Drops of NaOH to Titrate
1	Snail				
2	Organisms				
3	Org., 28°C				
4	Org. + Plant	——	——	——	
5	Plant, Light	——	——	——	
6	Plant, Dark	——	——	——	
7	Control	——	——	——	

5. Obtain three 5-cm long pieces of *Elodea* from one of the aquariums on the equipment table and transfer one piece to beakers "4," "5," and "6."

6. Obtain seven pieces of Parafilm® and one piece of aluminum foil from the equipment desk. Complete the four following instructions as quickly as possible: (1) cover the tops of each of the flasks with Parafilm®, (2) make note of the time, (3) place flask "3" in a 28°C water bath, and (4) completely cover flask "6" with foil so that no light reaches the plant.

7. After one hour has passed from the time that you covered the flasks with Parafilm®, proceed to the next point (8) of this experiment. In the meantime, obtain a clean fish net and seven clean 50-mL beakers from your teaching assistant and label them "1" through "7." To each beaker add four drops of the dilute phenolphthalein solution you will find on the equipment bench. Obtain a 1-mL pipette and a small flask containing about 25 mL of dilute sodium hydroxide (0.025 M) from the equipment bench.

8. When one hour has passed since you covered your experimental flasks with Parafilm®, retrieve your flask "3" from the 28°C water bath. Perform the next two points (9 and 10) of this experiment as quickly as possible.

9. Remove the Parafilm® cover from flask "1" and pour off 25 mL of the water (not the organism; use the fishnet to cover the top of the beaker) it contains into a beaker. Transfer this water to the correspondingly labeled 50-mL beaker to which you have already added four drops of phenolphthalein—use the increments on the side of the beaker to measure 25 mL as accurately as possible. Cover this 50-mL beaker with the Parafilm® you have just removed.

10. Repeat the last point (9) of this experiment in order for each of the remaining six experimental flasks.

11. Return the organisms in your experimental flasks to their original aquariums on the equipment bench. If you have grown so attached to your organisms that you would rather not part with them ask your teaching assistant if you can adopt them and they will help you make them a comfortable home at the end of the week's experiments.

12. Prepare to titrate with sodium hydroxide each of 50-mL beakers containing 25 mL of water and four drops of phenolphthalein. Titrate the water in each beaker individually by adding one drop of dilute sodium hydroxide followed by a gentle swirl to mix the solution. Wait at least five (5) seconds before continuing to titrate with sodium hydroxide and stop when you observe a permanent change in the color of the phenolphthalein (the solution's color will change to light pink and, if this color change persists for more than ten seconds, do not add any more sodium hydroxide). Record the number of drops of sodium hydroxide you had to add to each beaker in the fifth column of Table 8.1. Be sure to include a completed version of Table 8.1 in the results section of your laboratory report.

13. When preparing your laboratory report, subtract the number of drops of sodium hydroxide you had to use to titrate the negative control of your experiment (beaker "7") from the number of drops you used to titrate each of the other six samples. The remaining amounts of NaOH needed to reach the endpoint of the titrations is directly proportional to the amount of carbon dioxide present in the experimental flasks.

    In the results section of your laboratory report, be certain to report the relative metabolic rates (in units of drops of NaOH to titrate/g of body weight) of your experimental organisms in flasks "1," "2," and "3." Determine these relative rates by dividing the corrected number of drops of NaOH needed to titrate their water by the weight (in grams) of the organisms.

IMPORTANT QUESTIONS TO CONSIDER

Careful consideration of the following questions related to this set of experiments will not only insure that you understand the most important concepts of this laboratory but will also help you prepare for the lecture portion of the course. By addressing these questions (and others that you decide are important) in the conclusions section of your laboratory report, you will also assure that you receive full credit for that portion of your report.

- WHICH OF YOUR experimental beakers had the largest amount of carbon dioxide present in it at the end of the experiment and why?

- IS METABOLIC RATE related to the size and temperature of an organism's body?

- WAS THERE LESS carbon dioxide in the experimental beaker that contained your organism *and* a plant than the experimental beaker that contained just an organism? Why would you expect that to be the case?

- WHY WOULD THERE be a difference in the amount of carbon dioxide present in the beakers containing a plant kept in the light and a plant kept in the dark? Did you find a difference in your own experiment?

- WHAT FACTORS MAY have led to errors in your determining metabolic rates and carbon dioxide levels? How confident are you in your results?

EXPERIMENT 9:
CELL DIVISION AND DNA REPLICATION

OVERVIEW

In the two previous experiments you have indirectly studied the way in which cells carry on their metabolism and utilize carbon compounds as energy sources. From a biological perspective, there would be very little point in maintaining the complex enzymatic machinery that makes those processes possible if they did not ultimately preserve the organism's genetic material. The following set of experiments will allow you to examine in fine detail the most direct and effective means by which organisms accomplish that goal: cell division and DNA replication.

The cells of most of the organisms that are familiar to us are **diploid** meaning that they each contain two copies of the organism's genetic blueprint. An intricate series of events is played out every time those copies of genetic material are replicated and parceled out equally to two daughter cells.

The mechanisms used by cells to accomplish that process were established billions of years ago and are so tightly integrated, efficient and vitally important that they have hardly changed at all over the course of evolution. In this set of experiments, you will microscopically study the division of several kinds of cells. In the process you will make use of a window onto the ancient and fundamentally important process of cell division and DNA replication.

INTRODUCTION

The nucleus of a eukaryotic cell is its control center. The principle contents of each nucleus are a set of the organism's chromosomes upon which its genes reside. Experimental removal of a cell's nucleus provides a clear demonstration of the fundamental importance of this organelle and its contents; the existing machinery of a typical cell keeps it alive and functioning for several days but after less than a week the cell's movement becomes sluggish, feeding decreases and it ultimately dies.

Clearly then, the nucleus and the genetic blueprint it contains are absolutely essential to cells. It follows that when a cell divides, the information contained within its nucleus must be faithfully transmitted to both of the new cells that result. A simple splitting of the information into two halves would not be a satisfactory solution to this problem because neither of the two cells would have all the information they would need. Just as cutting the blueprint of a building in half would not allow two builders to construct two copies of the building, each of the new cells that result from cell division must be provided with a complete set of genetic instructions.

Rather than simply splitting its genetic information in half, cell division must involve a process of duplicating chromosomes *and* distributing those duplicates in an orderly fashion. The process by which the nuclei of eukaryotic cells divide to produce two new nuclei, each with the same number of chromosomes as the parental nucleus, is called **mitosis.** The separate process that results in the comparatively simple splitting of the cytoplasm and its contents is **cytokinesis.**

Cytologists, scientists who use microscopes to study cells, have divided the complete cycle of nuclear division into separate steps as a matter of convenience. Those steps are illustrated in BIO 1120's textbook and outlined in Table 9.1, but it is important to remember that the entire process of mitosis is actually a continuum without precisely defined beginnings and end points.

Table 9.1 The steps of mitosis.

Mitotic Step	Features
Interphase	Chromosomes are indistinct, nucleolus visible.
Early Prophase	Centrioles move apart, chromosomes appear as thin threads, nucleolus begin to disappear.
Late Prophase	Chromosomes thicken and move toward equator, nuclear membrane disappearing, nucleolus is no longer visible.
Metaphase	Nuclear membrane is no longer visible, centromeres are attached to spindles, chromosomes are line up on equator.
Anaphase	Chromatids seperate at centromeres, chromosomes move towards poles, cytokinesis begins.
Telophase	New nuclear membranes form, chromosomes become less distinct, cytikinesis nears completion.

Typical cells spend much more time in **interphase** than any other stage of their life cycle. During this time they perform the relatively mundane tasks of metabolism and growth as well as preparing for mitosis by replicating their DNA.

Prophase refers to the very earliest stages of mitosis. It too is something of a preparatory stage during which chromosomes first become visible as they coil into tighter and tighter helices. Individual chromosomes from a nucleus in late prophase can be seen with a microscope to consist of two separate strands called **chromatids.** Each chromatid is actually an all but identical copy of a single chromosome that was replicated during interphase. Careful examination with a microscope also reveals that there are actually *two* copies of each *pair* of chromatids as well—four copies of each chromosome are present within cells during this stage. Each pair of chromatids is tightly held together at a special region called a **centromere.**

During the point in time known as **metaphase,** chromosomes are neatly lined up on the equatorial plane of a dividing cell.

Examination with a microscope at this point reveals that each chromatid is attached to **spindle microtubules** which in turn are attached to a **centriole** located in each pole of the cell. Metaphase is considered complete when the centromeres holding each pair of chromatids uncouple and the chromatids begin to move apart.

The following step, **anaphase,** is the critical event for which all the previous stages were preparation. As the spindle microtubules to which they are attached shorten, each chromosome of every chromatid pair is drawn toward a centriole at an opposite pole of the cell. It is in this way that one copy of each of the cell's chromosomes is faithfully passed on to each of the two daughter cells that are made during cell division. Cells in early anaphase are readily recognized microscopically by the presence of two equal groups of highly condensed chromosomes only a short distance apart. In most organisms, cytokinesis also begins as anaphase proceeds.

Telophase marks the end of mitosis and is commonly described as the reverse of prophase. The two sets of chromosomes, now at opposite poles of the original cell, begin to decondense and become enveloped in new nuclear membranes. Cytokinesis rapidly completes the division of the cytoplasm and this stage is considered to be completed when the two daughter cells take on the interphase characteristics of the original single cell that gave rise to them. Notice that each daughter cell ultimately possesses *two* copies of each of its chromosomes and is diploid just as its mother was.

Sexual reproduction provided enormous evolutionary advantages to organisms that were capable of utilizing it early in the history of life on earth. It allowed the recombination of genetic traits with each new generation and provided means for natural selection to rapidly decrease the frequency of genes that reduced the fitness of organisms while simultaneously bringing advantageous genes together in particularly fit individuals. Sexual reproduction also posed organisms with a challenging hurdle—combining the genetic material of two organisms of opposite sex without creating an offspring that had double the number of chromosomes of its parents.

The answer to the problem of maintaining a constant number of chromosomes in sexually reproducing organisms came in the evolution of a special kind of nuclear division called **meiosis.** At a superficial level, meiosis is simply two rounds of mitotic division in rapid succession and before DNA replication can occur during an interphase (Figure 9.1).

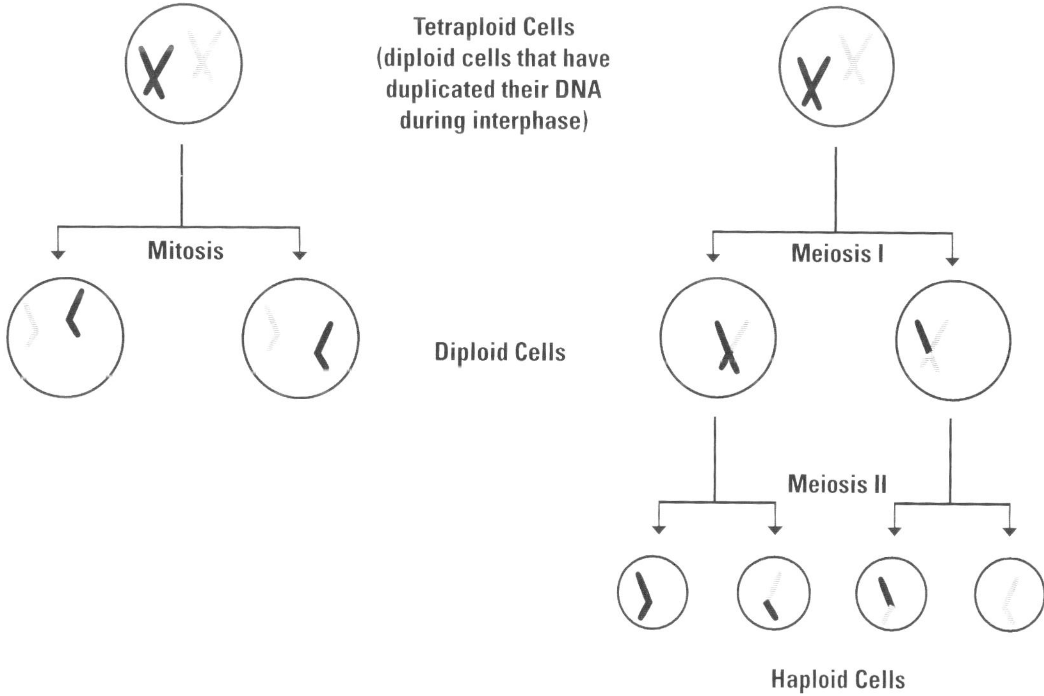

Figure 9.1 Schematic comparison of mitosis and meiosis. Each of the diploid cells receive two copies of a chromosome though they remain attached at the centromere in meiosis I but not in mitosis. Meiosis II halves chromosome number again so that haploid gametes are produced.

More formally though, meiosis in all sexually reproducing organisms involves two successive nuclear divisions that result in four new **haploid** cells (cells that contain only one copy of all of their genetic information unlike diploid cells which each have two copies). The first round of division gives rise to two diploid cells and reduces the number of copies of each chromosome after interphase from four to two. The second round of meiotic division separates chromatids and gives rise to a set of four haploid cells.

Many of the events in meiosis I resemble those of mitosis. The principal difference between the prophases of these processes is that members of each pair of chromosomes come to lie side by side at the cell's equator for metaphase. These matching (homologous) chromosomes often intertwine in meiosis I and are said to **synapse.** During this time, recombination between chromosomes often occurs resulting in exchanges of homologous pieces of the chromosomes as pictured in Figure 9.1. Another major difference between meiosis I and mitosis is that chromatids remain attached at their centromeres in anaphase of meiosis I.

The second division sequence of meiosis, meiosis II, is an essentially mitotic one. Still, the functional end result of meiosis II, a set of haploid cells called **gametes,** is different from mitosis. When two gametes (an egg and a sperm cell, for instance) from the same species combine their genetic material, a diploid zygote of that species is formed. In this process, two biologically important goals are achieved: (1) a new organism is created that possesses a sampling of the genetic characteristics of both its parents and, (2) the copy number of the organism's chromosomes has been maintained for another generation.

PROCEDURES

You will be utilizing the microscopes you used in week 4 of this semester as you perform the following experiments. Briefly review the material presented in "Experiment 4: Microscopes and Cellular Anatomy" of this laboratory manual pertaining to their operation and care prior to your arrival in the laboratory. Questions along those lines would be an excellent basis for a quiz at the beginning of this laboratory.

STEP ONE

1. Work in pairs. Obtain a microscope from the microscope cabinet and place it securely on your desk. Plug its electrical cord into one of the electrical outlets on your laboratory bench.

2. Obtain a mounted slide of a longitudinal section through an onion root tip from the supply bench. Examine the slide on low power (using the 4× objective lens) and locate and focus on the tip of the root.

    At the tip of onion roots is a tough section of dried cells (the **root cap**) that protects the tender root as it pushes through soil. Just behind the root cap is a zone of very active cell division and not far behind this region are cells that have fully differentiated into specialized cells such as those that transport water and nutrients to the rest of the plant.

3. Focus on the zone of actively dividing cells immediately behind the root cap using the low power objective lens, intermediate and then high power lenses. Survey this portion of the slide until you have found at least one example of each of the stages of the cell cycle: interphase, early prophase, late prophase, metaphase, anaphase, and telophase.

4. Refer to Table 9.1 to confirm your categorization of these dividing cells. Find the best examples of each stage of mitosis on your slide and make a rough sketch of them in Figure 9.2. Label the nucleus, nucleolus, chromosomes, mitotic spindle, and cell walls where appropriate on these sketches and be certain to include a completed version of Figure 9.2 with your answers to the questions at the end of this experiment.

5. Temporarily part company with your laboratory partner at this point. Decide who will perform the initial parts of Step Two and who will perform the initial parts of Step Three. While both of you should be familiar with the procedures and concepts involved in those steps, by working individually at this point you will be able to obtain results twice as quickly.

Interphase	Early Prophase

Late Prophase	Metaphase

Anaphase	Telophase

Figure 9.2 Sketches of the stages of mitosis. Label each nucleolus and cell wall as well as the chromosomes and mitotic spindles where visible in each of your sketches.

STEP TWO

Fruit flies are among the best understood of all eukaryotic organisms at a genetic level for a variety of reasons. One of the principal features that have made it such a valuable model organism however are the unusual chromosomes of its larval salivary glands.

The highest priorities of fruit fly larvae are all related to food—particularly its digestion. The salivary glands of fly larvae are remarkably active, and they grow largely through an increase in cell mass and volume rather than an increase in the number of cells. As the size of the nuclei of these giant cells continues to increase, the chromosomes within them continue to duplicate again and again but without accompanying nuclear divisions. The extra chromosomes that result probably support increased levels and control of activity in these large cells.

Typically, chromosome copies only pair during meiosis. However, in these unusual larval salivary glands not only do homologous chromosomes pair, but thousands of copies of the duplicated chromosomes are tightly held together side by side. The result is a structure known as a **polytene chromosome** that can be easily seen with the help of a microscope.

Polytene chromosomes provide an unusual advantage in the assigning of genes to specific regions of chromosomes. When stained with aceto-orcein, they take on a dramatic striped appearance with numerous dark staining expanded regions called **bands** (or puffs). The width and arrangement of these bands are so characteristic that they can be used to uniquely identify even small segments of chromosomes. The tools of molecular biology have been used with great success to assign specific genes to many of these bands.

In this step of the experiment, you will prepare chromosome **squashes** of fruit fly polytene chromosomes and observe these unusual and useful structures with the help of a microscope.

1. Working alone, move to one of the six dissecting microscopes at the laboratory bench in the front of the laboratory when one is available. Place a single drop of 0.7% saline solution on one of the clean slides that you will find next to the microscope. Use a dissecting needle to transfer a single fruit fly larva from the bottle next to the microscope to the drop of saline on the slide.

2. The operation of a dissecting microscope is very similar to that of the microscopes you have been using. Spend a few minutes becoming familiar with this new microscope by examining the fruit fly larva and manipulating it with dissecting needles.

3. While looking through the dissecting microscope, use a pair of dissecting needles to pull the head off of the fruit fly larva. This decapitation is most easily accomplished by: (1) placing one needle in the middle of the larva and the other just behind the relatively hard mouthparts of the head, (2) pulling the needle at the head with a jerk and then relaxing, and (3) watching as two salivary glands on either side of a tube-like gut are slowly pushed out of the body.

4. The large individual cells of the salivary glands should be clear and grape-like in appearance. Try to tease off the opaque pieces of fat that are likely to be attached to the salivary glands.

5. Place a drop of aceto-orcein stain on a spot on your slide that is well-removed from the drop of saline. Aceto-orcein stains DNA. Your skin has DNA in it. Don't stain your skin with aceto-orcein.

6. Use the dissecting needles to separate the salivary glands from the rest of the body of the larva and transfer them to the drop of stain. Use the dissecting microscope to confirm that the transfer was successful and return to your laboratory bench.

7. After the salivary glands have stained for ten minutes (don't let the stain dry-out while you wait), place a coverslip on the preparation. Place the slide and coverslip between the folds of a paper towel and press down on the coverslip firmly with the eraser end of a pencil. Pressure applied with the eraser end of a pencil can also be used to "squash" the stained salivary glands.

8. Locate chromosomes on the slide you have just made with the 4× objective lens of the microscope at your laboratory bench. When you find a clustering of them, make a sketch of what you see in Figure 9.3. Invite your laboratory partner to view the polytene chromosomes you have found and make a sketch of them in Figure 9.3 as well so that you can both include one with your answers to the questions at the end of this report.

    Both you and your partner should also record how many different chromosomes seem to be clustered together in your "polytene spreads"—that number corresponds to the haploid chromosome number of fruit flies since homologous chromosomes are so tightly associated when in the polytene state.

Figure 9.3 Sketch of the giant polytene chromosomes of a fruit fly salivary gland cell.

STEP THREE

Sordaria fimicola is a fungus that spends most of its life as a mass of haploid cells associated in filaments known as a **mycelium.** When this fungus lives in a rich environment, cells within filaments from two different mating types fuse. Single-celled, diploid zygotes result and each is encased in an **ascus** (plural, asci). Within the asci each single-celled zygote undergoes a single round of meiotic division and gives rise to four haploid cells. These haploid cells then undergo a single round of mitotic division and give rise to a total of eight haploid **ascospores** with thick cell walls in each ascus. Many of these rod-shaped asci, each with eight ascospores, are held together in *Sordaria* fruiting bodies.

Conveniently, from the perspective of geneticists, all the ascospores within an ascus remain ordered in the same position that their chromosomes were aligned in during the first round of meiosis (meiosis I). This unique sequence of events allows a simple detection of the occurrence of crossing over that occurs during meiosis in this fungus. A single gene with two alleles governs whether ascospores are black *(B⁺)* or tan *(B⁻)*. Remember that *Sordaria* spends most of its life cycle in a haploid state and that the genes present in the ascospores are all normally expressed. As a result, the order of ascospore colors within an ascus can be used to detect the frequency that these alleles were exchanged (crossing over had occurred) between chromosomes during meiosis (Figure 9.4).

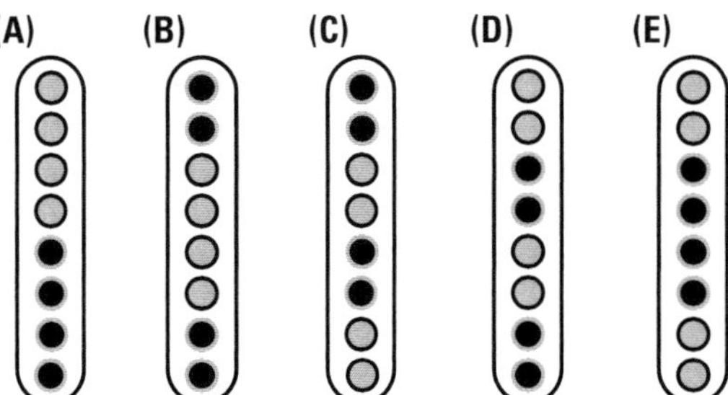

Figure 9.4 Possible arrangements of colored ascospores within fungus asci. A single gene with two alleles codes for either black or tan asci. No recombination has occurred in **(A)**, but at least one must have taken place to give rise to asci looking like **(B)**, **(C)**, **(D)**, and **(E)**.

The greater the distance between two points on a linear chromosome, the greater the likelihood that crossing over will occur between them. If two spots (**loci,** plural of **locus**) on a chromosome are close together, relatively few recombinations of those points will be observed after each meiosis and the points are said to be tightly linked. By observing the recombination frequencies of genes on the same chromosome, then it is possible to produce a **linkage map** that reflects their positions relative to each other.

In this step of the experiment, you will observe living cultures of crosses that the prep room personnel have made two weeks in advance between black *(B⁺)* and tan *(B⁻)* colored *Sordaria*. By determining the abundance of the various arrangements of ascospore colors within asci, you will be able to estimate the recombination frequency of the gene for ascospore color and the centromere of the chromosome on which that gene resides.

1. Working alone, obtain a clean slide, coverslip, toothpick and *Sordaria* culture plate from the equipment bench and return to your laboratory bench. Place a single drop of water on the center of your slide.

2. Open the lid of *Sordaria* culture and use a toothpick to remove a very small amount of material on the **surface** (do not dig into the agar of the plate) of the culture where you can see that strains of the two different colors have grown together. The best fruiting bodies are often found close to the outer rim of the culture dish. Without setting the toothpick down, return the lid to the *Sordaria* culture.

3. Touch the toothpick to the drop of water on your slide to move asci onto it. Cover the drop of water with the coverslip and tap it *lightly* with the eraser end of a pencil to break open and flatten out the asci. Return the *Sordaria* culture to the equipment bench so that others in your laboratory section can harvest asci from it as well.

4. Place the slide you have prepared on the stage of your microscope and scan the drop for asci using the low or medium power objective lenses (4× or 10×). When you locate clusters of asci focus on them—switch to the highest power objective lens at this point if you are more comfortable viewing the asci in that way.

5. Confirm that asci fall into one of three categories in which ascospores are: (1) all of the same color (the zygote that gave rise to them was formed by fusion of mycelium of the same strain), (2) four in a row of one color and **four in a row of the other** (no recombination at the color locus occurred in the zygote, see Figure 9.4A) and (3) **two of one color** next to two of the other color next to two of the first color next to two of the other color or four in a row of one color flanked on both sides by **two in a row of the other color** (at least one crossover occurred in the zygote, see Figure 9.4B through E).

6. Ignore all asci that can be classified as class I (all ascospores are the same color) and count those that belong to classes II and III. Continue counting and recording the relative numbers of each class in Table 9.2 until you have observed at least fifty (50) asci in which recombination could have occurred (classes II and III). Share your data with your laboratory partner so that you will both be able to include a completed version of Table 9.2 with your answers to the questions at the end of this experiment and invite them to observe the asci you have examined.

Table 9.2 Relative numbers of recombinant and non-recombinat asci.

	Class II	Class III
Number Observed		

7. In preparing your laboratory reports, both you and your partner should separately use the data in Table 9.2 to determine the frequency of recombination between the gene for ascospore color and the centromere (the number of class III events divided by the sum of the number of class II and class III events).

8. An arbitrary unit of measure, the **map unit,** is commonly used to describe the distance between two loci on a chromosome. One map unit is equal to a 1% recombination frequency. In *Sordaria*, since only four of the eight ascospores in each ascus are the direct result of meiotic crossing over (the other four result from mitotic divisions), the number of map units between two points on a chromosome is equal to the frequency of recombination seen in asci divided by two. Calculate the distance in map units between the centromere and the gene for asci color and include this with your answers to the questions at the end of this experiment.

IMPORTANT QUESTIONS TO CONSIDER

Careful consideration of the following questions related to this set of experiments will not only insure that you understand the most important concepts of this laboratory but will also help you prepare for the lecture portion of the course. By addressing these questions (and others that you decide are important) in the conclusions section of your laboratory report, you will also assure that you receive full credit for that portion of your report.

- WHICH OF THE phases of mitosis did you see the most and fewest examples of in your microscopic examination of onion root tips?

- WHAT CONCLUSIONS CAN you draw about the relative length of each of the stages of mitosis from your examination of onion root tips?

- WHAT ONION CELLS would you have to look at to examine meiosis as well as mitosis?

- BASED ON YOUR examination of polytene chromosomes in fruit flies, what is the normal number of chromosomes present in most fruit fly cells?

- HOW DIFFICULT WAS it to distinguish between two stretches of polytene chromosome by looking at differences in their banding patterns?

- WOULD YOU HAVE to look at more or less asci to observe the same number of recombination events between the gene for ascospore color and the centromere if you used a gene that was closer to the centromere?

EXPERIMENT 10:
HUMAN AND POPULATION GENETICS

"I'm one of those people you hate because of genetics. It's the truth." Brad Pitt

OVERVIEW

For thousands of years, all manner of people have been intensely interested in the physical traits that apparently run in families and the way that those traits gradually change in subsequent generations. It has only been within the last hundred years that geneticists and population geneticists, respectively, have provided satisfying answers to these naturally fascinating questions.

Modern genetics began with Gregor Mendel's studies of garden peas in the late nineteenth century and it was only in the 1950s that DNA was recognized as the chemical material within our cells that was responsible for much of our inheritance. It has since become possible to trace seemingly insignificant changes in the order of the nucleotides within our DNA through the dramatic affects they have on individuals. Perhaps even more importantly, it is also possible to do the reverse and predict which individuals are most likely to possess certain traits due to the nucleotide sequence of their genes.

During the course of this experiment, you will examine a number of different genetic traits in yourself, your family, and your fellow students. Through a study of these characteristics at levels ranging from the molecular to the anatomical, you will be able to determine how they are passed on from generation to generation and the forces that cause the frequency of those traits to change over the course of time.

INTRODUCTION

Not many people would readily admit to finding fruit flies, pea plants, bacteria, or yeast more interesting than people. Geneticists are no exception.

Still, many geneticists spend their careers studying organisms other than humans because humans just are not good model organisms for genetic experiments. Fortunately, the insights gleaned from other organisms are generally applicable to us as well. For instance, it was Gregor Mendel's work with garden peas in the late nineteenth century that suggested some genetic traits suppress the affects of others and the concepts of **dominance** and **recessiveness** have proven to be extremely important in the study of human genetics as well.

Aside from our long generation time and the difficulties in arranging crosses (not to mention F_1 crosses!) between individuals with particular traits, the biggest problem associated with studying genetic traits in humans is the fact that it is unethical to deliberately mutagenize people. Nonetheless, years of difficult study of genetic defects that already exist in individuals have revealed that single genes control many biochemical traits in humans. Much of the work in this field depends upon sophisticated analyses of **pedigrees** such as the one shown in Figure 10.1. Much can be learned simply from the pattern of inheritance seen in families that have individuals affected with particular genetic diseases. It is the finding of correlations between molecular markers and the diseased state on pedigrees however that has been a particularly important step in the characterization of many human genes.

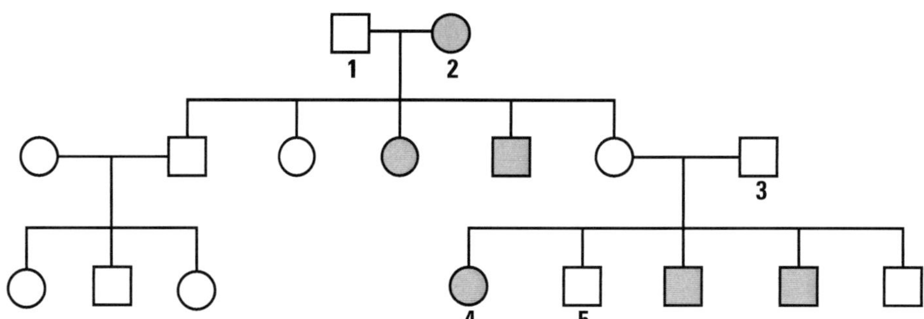

Figure 10.1 A pedigree illustrating the transmission of a recessive trait such as albinism in a family. By convention, open symbols are used to represent individuals with a normal phenotype and filled symbols represent affect individuals. Males are usually symbolized by squares and females by circles. A "T" connecting the symbols for a man and a woman represents their union and offspring. Pedigrees such as this one allow an unambiguous determination of the genotypes of individuals such as those numbered 1 through 4. If individual 5 marries an albino woman, what is the chance that their children will also be albinos?

The list grows longer almost every day, but some familiar examples of human diseases caused by defects in single genes are sickle cell anemia, phenylketonuria (PKU), albinism, cystic fibrosis, some forms of colon cancer and Alzheimer's disease. As the genetic basis of each of these diseases has become better understood, tests have been designed that allow prospective parents to know the risk of passing these traits on to children. And, in more and more cases, these insights have allowed treatments and even cures to be developed that virtually eliminate the threat posed to affected individuals.

Most physical traits in humans are also genetically controlled. Several genes play a role in determining eye color while several others are involved in determining hair color, height, and many other traits that make us physically different. When more than one gene plays a role in determining a characteristic, that trait is said to involve **polygenic** inheritance. However, like flower color in Mendel's garden peas, many human characteristics are also known to be controlled by single genes.

An examination of Mendel's cross of garden peas with flowers of different color perfectly parallels discussions of most of the human traits you will examine in this set of experiments. Mendel's earliest genetic inquiries involved two forms of the gene for flower color: one that gave rise to red flowers and the other to white. When a gene exists in more than one form in this way, the different forms are referred to as **alleles.**

Mendel soon learned that the gene for red flowers was dominant to the recessive one for white flowers. He was the first to begin the convention of assigning single letters to designate genes and using the capital of the letter to represent the dominant form of the gene and lower case to represent its recessive allele. In the case of flower color, he used C to symbolize the gene for red color and to c symbolize the gene for white flower color.

Remember that garden peas, like people and all other eukaryotes, are diploid organisms meaning that they have two doses of each of their genes. Thus, the cells of a pea plant may be **homozygous** and contain two copies of the gene for red flowers *(C/C)*, or two copies of the gene for white flowers *(c/c)*, or they may be **heterozygous** and contain one copy of each *(C/c)*.

All of the white-flowered plants that Mendel studied had to be homozygous recessive *(c/c)*. However, it was not possible to determine by visual inspection of the physical characteristics of a red-flowered pea plant (its **phenotype**). A red-flowering plant could either be a homozygous dominant *(C/C)* or a heterozygote *(C/c)* because both **genotypes** will have red flowers. Only genetic crosses and an examination of the resulting progeny could distinguish between the two types of red-flowering plants. For instance, if a cross between a red- and white-flowering plant resulted in an equal mixture of red- and white flowering progeny, it could be determined that the red-flowering parent was a heterozygote *(C/c)* (Figure 10.2A). If the same mating resulted in progeny that were exclusively red-flowering however, the red-flowering parent must have been a homozygous dominant *(C/C)* (Figure 10.2B). Until such crosses are performed, individuals that display dominant traits such as red flower color are often represented as *C/_* to reflect that uncertainty.

Precisely the same type of reasoning can be applied to human physical traits such as albinism (observed only in homozygous recessive individuals) as depicted in Figure 10.1.

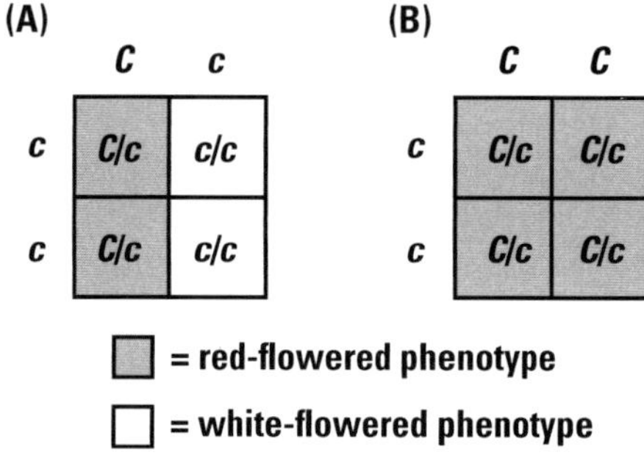

Figure 10.2 (A) A Punnett square representing a cross between a *c/c* individual (vertical) and a *C/c* (horizontal) individual. The ratios of genotypes (heterozygotes to homozygous recessives) and phenotypes (red-flowered to white-flowered) in this case are the same (1:1). **(B)** A Punnett square representing a cross between a *c/c* and a *C/C* individual. All progeny have the same genotype (they are heterozygous) and phenotype (red flowers).

Darwin's most important insight into evolution was that **natural selection** acts to change the relative frequencies of alleles within populations. Individuals that are burdened with genes that gave rise to undesirable physical traits simply do not pass those genes on to progeny as effectively as individuals with more advantageous genes are able to pass on theirs. Populations evolve, individuals do not.

The insights of Mendel and Darwin laid the groundwork for G. H. Hardy (an English mathematician) and W. Weinberg (a German physician) to independently formulate in 1908 what has become known as the **Hardy-Weinberg law.** In its simplest form, the Hardy-Weinberg law maintains that in the *absence* of natural selection and appreciable levels of mutation, the relative frequency of alleles within a population as a whole *will not change.*

In other words, if there are only two alleles in a population for a given gene and the frequency of allele C is 0.6 and the frequency of allele c is 0.4, the frequency of those two alleles will remain 0.6 and 0.4 for an indefinite number of generations as long as natural selection does not favor one allele over the other and mutations to the gene are random and rare. One of the most useful extensions of this observation is that allele frequencies within a population can be expressed as the binomial expression $(p + q)^2$ where p is the frequency of the dominant allele (C, or 0.6) and q is the frequency of the recessive allele (c, or 0.4). Expansion of this expression yields the formula for the "Hardy-Weinberg equilibrium" state or:

$$p^2 + 2pq + q^2 = 1. \tag{10.1}$$

The equation for Hardy-Weinberg equilibrium (9.1) is remarkably powerful. For example, if it has been determined that 16% of all the pea plants in a particular garden are white flowering, then the value of q^2 for this garden is 0.16. If $q^2 = 0.16$ then the square root of q^2 (more easily referred to as q) is 0.40. If $q = 0.40$ then p must equal 0.60 since the total of all the alleles in a population ($p + q$) must equal 1. If $p = 0.60$ then the fraction of plants that are homozygous dominants or $p^2 = 0.36$ and the fraction that are heterozygotes is $2pq = 2 \times 0.60 \times 0.40 = 0.48$.

Don't let the little bit of algebra intimidate you—look at what we have just learned without having to perform and wait for the results of a single cross between a red and white flowering pea plant from this garden! Simply by determining the relative number of white flowering plants we have been able to infer the relative numbers of homozygous dominant and heterozygote plants in the population.

There is also an interesting lesson for those of you who will become tyrants and have an eye for creating a master race when you graduate—in most naturally occurring populations heterozygotes harbor a large fraction of both the dominant and recessive alleles. In the example above the number of recessive alleles hidden in heterozygotes is almost equal to the number present in homozygous recessive individuals—it would take a very large number of generations to totally eliminate an undesirable recessive trait if all you did was remove the individuals that displayed it. That insight also explains why so many genetic diseases coded by recessive genes linger in human populations for so long.

PROCEDURES

In this course of the first step of this experiment, you will examine several of your own physical characteristics that are known to be controlled by single genes within the human genome. Before arriving in the laboratory, you should become acquainted with these traits and, if possible, look for them in your parents and siblings. The traits you will be examining are as follows:

- **Tongue rolling:** The dominant R allele for this gene confers the ability to roll your tongue into a "U" shape while those who are homozygous for the recessive r allele cannot.

- **Widow's peak:** The dominant W allele for this gene causes the hairline above the forehead to be "V" shaped while those who are homozygous for the recessive w allele have straight hairlines.

- **Thumb crossing:** Most people interlace their fingers exactly the same way every time they hold their hands together. Interestingly, this (and the related trait of arm crossing) may be a heritable trait. The dominant C allele compels individuals to place their left thumb on top of the right thumb when they interlace their fingers, while the recessive c allele compels individuals to place their right thumb on the top of the left thumb when they interlace their fingers.

- **Attached ear lobes:** The E allele for unattached earlobes is dominant to the recessive e allele for attached earlobes.

- **Dimpled chin:** Individuals with the dominant allele for this gene, D, have a cleft in their chins while those who are homozygous for the recessive d allele do not.

- **Mid-digital hair:** The dominant M allele for this gene causes individuals to have hair on one or more of their fingers between the first and second joint (not the part closest to where the fingers are joined to the base of the hand). Those who are homozygous for the recessive m allele have a complete absence of hair.

- **PTC-tasting:** A dominant allele, $T,$ confers upon a small number of individuals the ability to detect a bitter taste when they place a small amount of the chemical phenylthiocarbamide (PTC) on their tongue. Those who are homozygous for the recessive t allele will always have to wonder what they are missing.

- **Lactose tolerance:** A single gene codes for the enzyme that breaks milk sugar (lactose) into its component monosaccharides and this gene is functional in virtually all individuals. A second regulatory gene however causes this metabolic gene to be turned off in most adult primates. Individuals who are homozygous for the recessive *l* allele of this gene do not make the regulatory protein and can digest lactose throughout their adult lives. Individuals with the dominant allele *L* will know that they are not "lactose tolerant."

- **"Asparagus urine":** Individuals who have a dominant *O* allele of this gene are capable of smelling metabolic byproducts of asparagusic acid which is particularly abundant in asparagus. Those byproducts build up quickly (like in 15 minutes) in urine after people eat asparagus, but only those with a dominant version of the gene can smell them. Until recently it was thought that some humans metabolize asparagus differently than others such that some people excreted odorous urine after eating asparagus and others did not. However, we now know that producing odorous urine from asparagus is a universal human characteristic—it is the ability to detect the smell that is unusual!

The amino acid asparagine gets its name from asparagus which is particularly rich in that component of proteins. At the beginning of this lab, you will have the opportunity to take an asparagine supplement. If you take it, you will be able to tell by the end of the laboratory period if you are capable of smelling "asparagus urine." If you have any doubt by the end of the experiment, you can safely count yourself among those that do not have at least one copy of the dominant *O* allele. Even one piece of asparagus should be enough to replicate this exercise at home any time you like.

STEP ONE

1. For each of the traits listed above, determine your phenotype and circle the appropriate entries in Table 10.1 below. Test strips containing the chemical PTC will be available for you to test that trait—place a piece of PTC paper on your tongue and let it rest there for about ten seconds (if you have any doubt about your ability to taste it you are a non-taster of PTC).

   Include Table 10.1 in the results section of your laboratory report.

Table 10.1 Human genetic traits seen in yourself.

Characteristic	Your Phenotype	Your Possible Genotypes
Tongue Rolling	"U" Shaped/Flat	*R/R, R/r, r/r*
Widow's Peak	"V" Shaped/Straight	*W/W, W/w, w/w*
Thumb Crossing	Left on Top/Right on Top	*C/C, C/c, c/c*
Attached Ear Lobes	Unattached/Attached	*E/E, E/e, e/e*
Dimpled Chin	Dimpled/Un-dimpled	*D/D, D/d, d/d*
Mid-digital Hair	Hairy Fingers/Bald Fingers	*M/M, M/m, m/m*
PTC-tasting	Taster/Non-taster	*T/T, T/t, t/t*
Lactose Tolerance	Intolerant/Tolerant	*L/L, L/l, l/l*
"Asparagus Urine" Smelling	Smelling/Non-smelling	*O/O, O/o, o/o*

2. Report your results to your teaching assistant. Make a note of the total number of students in your section. Fill in the second and third columns of Table 10.2 with the cumulative data for your laboratory.

Use the total number of students in your section to determine the frequency of each phenotype and complete the remainder of Table 10.2. Be sure to include the completed version of Table 10.2 in the results section of your laboratory report.

Table 10.2 Human genetic traits seen in your laboratory's population.

Characteristic	Students with Dominant Phenotype	Students with Recessive Phenotypes	Frequency of Dominant Phenotype	Frequency of Recessive Phenotype
Tongue Rolling				
Widow's Peak				
Thumb Crossing				
Attached Ear Lobes				
Dimpled Chin				
Mid-digital Hair				
PTC-tasting				
Lactose Tolerance				
"Asparagus Urine" Smelling				

3. After compiling the numbers in Table 10.2, your teaching assistant will ask everyone in your laboratory section to stand. Your teaching assistant will tell you their own phenotype for each of the traits you have examined in the same order in which they appear in Table 10.1 and Table 10.2. Sit down when you hear a phenotype different than your own. Report in the results section of your laboratory report how many traits needed to be considered before you could be differentiated from your teaching assistant and how many needed to be considered before everyone in your section sat down.

4. The remainder of Step One (points 4, 5, and 6) can be completed at home.

   From the frequency of individuals displaying recessive traits in your laboratory section (q^2), determine the frequency of the recessive allele (q) for each of the traits you have studied. Record these frequencies in Table 10.3.

5. Knowing that the total frequency of all alleles must always equal 1, subtract the frequencies for q from 1 and record the value for p in Table 10.3.

6. Use the values you have calculated for p and q to infer the number of heterozygotes ($2pq$) for each trait in your laboratory section and record these results in Table 10.3. From the value you have calculated for p, determine the frequency of homozygous dominant individuals (p^2) for each trait and record the values in Table 10.3. Be sure to include the completed version of Table 10.3 in the results section of your laboratory report.

Table 10.3 Hardy-Weinberg equilibrium predictions for a population.

Characteristic	q	p	$2pq$	p^2
Tongue Rolling				
Widow's Peak				
Thumb Crossing				
Attached Ear Lobes				
Dimpled Chin				
Mid-digital Hair				
PTC-tasting				
Lactose Tolerance				
"Asparagus Urine" Smelling				

STEP TWO

All of the human traits you have studied in this experiment to this point are totally the result of the influence of single genes with only two possible alleles in your genome. A large majority of human characteristics however are actually influenced by many genes as well as by our environment. Height, for instance, is the result of the cumulative effect of at least seven different genes and is also heavily influenced by diet (general improvements in which have resulted in a steady and substantial increase in the height of human adults for the past 200 years).

Another characteristic that is determined both by polygenic inheritance and the environment are fingerprints. The epidermal ridges on our fingers, palms, soles, and toes are all formed late in our embryonic development when sacs of fluid covering those areas collapse just before birth. The way in which those fluid sacs collapse and the ridges they leave us with the rest of our lives depends largely on what, when and where we bump in to things as embryos. As a result, even twins that have the same genetic make-up can still be distinguished by their fingerprints.

Still, even though the environment plays a role in the formation of our epidermal ridges, a number of genes also influence that characteristic. The fraction of the phenotypic variance of a trait such as fingerprint patterns that is due to genetic differences is known as its **heritability (H).**

The heritability of a trait is that proportion of its variance that is caused by genetic variance. If the heritability of a trait is 100% such as it is for the eight characteristics you examined in the first part of this experiment then $H = 1.0$. On the other hand, if all the variation observed is due to environmental factors then $H = 0.0$.

If V_T is the total phenotypic variance of a trait observed between two or more samples, V_G is the fraction of the phenotypic variance that is due to genetic differences among individuals and VE is the fraction of variance that is due to the environment, then:

$$V_T = V_G + V_E \tag{10.2}$$

and, by definition:

$$\text{heritability} = H = V_G / V_T. \tag{10.3}$$

Determining the phenotypic variance of a trait is not difficult. First, calculate the mean value for the trait in the group being considered, then determine the differences between each value and the mean. The average of the square of these differences is the variance.

The genetic contribution to the fingers of your left and right hands should be exactly the same since virtually every cell in our bodies carries a nearly perfect copy of the same set of genes. As a result, any variance in the total number of ridges (**total ridge count** or **TRC**) on the fingers of your left and right hands must be due to environmental influences and can be considered to be V_E for that trait.

The variance in TRC in a population the size of your laboratory section is a relatively good measure of the total variance (V_T) for that characteristic. From Equation (10.2) it follows that the genetic contribution (V_G) to TRC is equal to $V_T - V_E$ and from Equation (10.3) we know that the heritability of this trait can be determined simply by dividing V_G by V_T.

In this step of the experiment, you will take a set of your fingerprints and determine the TRC for the fingers on your left and right hands. By comparing the variance between the TRCs on each of your hands and those of your class in general you will effectively determine the heritability of this trait.

1. Press one of your fingers firmly on an ink pad and make sure that you have evenly covered the sides of your finger with ink.

2. Gently and steadily roll your ink covered finger on the appropriate box on Figure 10.3. Be certain to also get an impression of the sides of your finger.

3. Repeat points 1 and 2 using different fingers until you have a complete set of your fingerprints. Include a completed version of Figure 10.3 in your laboratory report.

4. On most of your fingerprints, you will find one or two **triradi** (or "deltas"). A triradius is a point at which three groups of ridges from three different directions intersect. Dermatoglyphographers (people who study fingerprints) determine a characteristic ridge count for fingerprints by counting the number of ridges from the center of the pattern to a triradius. If more than one triradius is present, the count for the highest number of ridges is used. If no triradius is present (a relatively rare occurrence), a ridge count of zero is reported.

    Use a magnifying glass to determine the total ridge count (TRC) for each of your fingerprints. Record your results in Table 10.4.

5. Sum up the TRC for fingers on your left hand and for each of the fingers on your right hand. Include those values in the results section of your laboratory report and report them to your teaching assistant.

6. Determine the variance in the total ridge count on the fingers of your left and right hands that is caused by environmental influences (V_E). Obtain from your teaching assistant a value for V_T that was determined from the total ridge counts of all the members of your laboratory section. Use these values to determine the variance of this characteristic that is due to genetics (V_G). Calculate the heritability (H) of fingerprints and report all of these values in the results section of your laboratory report.

Table 10.4 Total ridge counts (TRC) for each of ten fingerprints.

Finger	L–1	L–2	L–3	L–4	L–5	R–1	R–2	R–3	R–4	R–5
TRC										

L–1

R–1

L–2

R–2

L–3

R–3

L–4

R–4

L–5

R–5

Figure 10.3 Fingerprint set. Boxes correspond to each of your ten digits with L–1 being your left thumb, L–2 your left index finger and so on.

IMPORTANT QUESTIONS TO CONSIDER

Careful consideration of the following questions related to this set of experiments will not only insure that you understand the most important concepts of this laboratory but will also help you prepare for quizzes and the lecture portion of the course. By addressing these questions (and others that you decide are important) in the conclusions section of your laboratory report, you will also assure that you receive full credit for that portion of your report.

- WHICH OF YOUR genotypes would you be able to better determine if you considered the phenotypes of your parents and siblings?

- WHY DON'T RECESSIVE traits always eventually disappear from populations?

- WHAT FRACTION OF recessive alleles are "hidden" in heterozygotes for each of the eight single-gene traits that you studied?

- HOW MUCH WOULD the frequency of the recessive gene for the inability to taste PTC be reduced if a tyrant eliminated all those who could not taste PTC from three successive generations before they could pass on their recessive alleles?*

- WHY MIGHT YOU expect the locus that governs lactose tolerance/intolerance to not be in Hardy-Weinberg equilibrium?

- WHAT MAY HAVE led to errors in your determining your total ridge count and how confident are you in the reliability of the heritability for this trait that you calculated?

- IQ IS ANOTHER characteristic that is apparently heavily influenced by both genes and environment. What other human characteristics are likely to fall into this class?

An amazing amount of information about the traits you have studied in this lab (as well as more than anyone could reasonably be expected to imagine) are catalogued at "On-line Mendelian Inheritance in Man." This resource is famous to those in the trade of human genetics simply as OMIM and can be found on the web at: *www.ncbi.nlm.nih.gov/omim/*.

*This would be an example of what population geneticists would call "harsh artificial selection." The frequency of a recessive allele is reduced much more quickly when it is large than when it is small because recessive alleles are mostly hidden in heterozygotes when their frequency is low—this is one of the reasons that some genes responsible for serious human diseases like cystic fibrosis persist in human populations indefinitely. Also note how much easier it is to eliminate a dominant allele—harsh artificial selection can eliminate dominant alleles in just one generation.

EXPERIMENT 11:
DNA ELECTROPHORESIS
AND QUANTITATION

"Trying to read our DNA is like trying to understand software code—with only 90% of the code riddled with errors. It's very difficult in that case to understand and predict what that software code is going to do." Elon Musk

OVERVIEW

Molecules are unimaginably small. The double-stranded molecular structure of DNA is so fine that a DNA molecule long enough to circle Earth at the equator would weigh less than a grain of fine sand. The same length of cotton thread would weigh about 600 pounds.

Still, molecular biologists need to be able to isolate, manipulate and characterize DNA as a routine part of their studies. That generally means separating large numbers of DNA molecules on the basis of their size and determining just how much DNA is present in various stages of experiments.

In this laboratory you will be performing experiments to characterize and quantify nucleic acids. You will find that the quantities of reagents used in molecular biology experiments are indeed very small and that very often it is not even possible to see the DNA molecules of interest until after special stains have been used to help us see where large numbers of them have moved during the course of a series of procedures.

INTRODUCTION

The movement of charged molecules through a matrix or solvent driven by an electric field is called **electrophoresis.** This technique is routinely used by molecular biologists to: determine the size of DNA and RNA molecules; assess the purity of isolated DNA and RNA molecules; and estimate concentrations of DNA and RNA samples. In zonal methods like electrophoresis molecules are placed in a thin layer or zone of a matrix and then forced to move through the matrix. The matrix minimizes the effects of convection and diffusion and can be any of a wide variety of materials such as paper, cellulose, silica gel, agarose, starch, or polyacrylamide. All the matrices used for electrophoresis ultimately also act as sieves by impeding the movement of some molecules of interest more than others. By carefully selecting an appropriate electrophoretic matrix, molecules can be resolved due to differences in a variety of different properties such as their size, charge or isoelectric point. An electrophoretic setup is shown in Figure 11.1.

The rate and direction a molecule moves through an electrical field, its **electrophoretic mobility,** depends upon the biochemical properties of the molecule (its net charge, charge density, size, and shape) and the properties of the electrophoretic medium and buffer being used (their ionic strength, temperature and viscosity). Charged molecules are naturally attracted to an electrical pole with charge opposite to theirs when placed in an electrical field. DNA and RNA molecules are typically strongly negatively charged due to the phosphates of their phosphodiester backbones. Proteins in their native states may have a positive or negative net charge depending on their amino acid composition and the pH of the surrounding media. If a protein has more acidic amino acids (glutamic and aspartic acids) than basic amino acids (lysine, arginine, and histidine), then the protein will have a net negative charge at neutral pH. However, the pH of the surrounding media can cause protons to be added or removed from amino acids change a protein's net charge. But, the highly acidic nature of every one of their phosphate groups make DNA and RNA molecules negatively charged in cells and typical laboratory experiments.

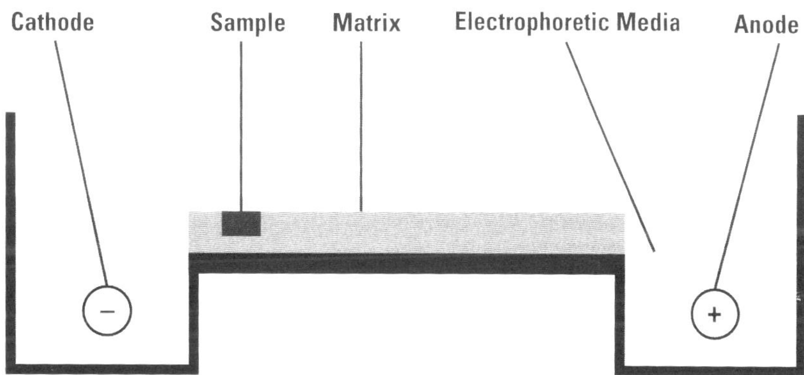

Figure 11.1 Diagram of an electrophoresis set-up.

Electrophoresis is most often used to separate DNA (and sometimes RNA and protein) molecules on the basis of their size. Electrophoretic matrices with gel-like consistencies form small pores that sieve the molecules by allowing smaller molecules to move more quickly than larger ones. Nucleic acids are usually electrophoresed through polyacrylamide or agarose gels. Agarose is an inert polymer of *D*-galactose and 3,6-anhydro-*L*-galactose extracted from seaweed. The size of pores in agarose gels is determined by the concentration of agarose used to make the gel (higher concentrations have smaller pore sizes than lower concentrations) and molecules ranging in size from 200 to 50,000 base pairs in length can be routinely separated. Polyacrylamide gels are usually used for finer scale separations of smaller DNA molecules and can separate molecules that differ in length by a single nucleotide.

The molecular size of nucleic acids is affected by both their mass (their length in nucleotides) and their three dimensional shapes. Linear DNA molecules generally "snake through" such that their mobility is related to the probability that one of their ends is near a pore opening. The longer a DNA molecule is, the less likely it becomes that one of its terminal nucleotides will be the first to encounter a pore opening. As a result, the $\log_{10}$ of the base pair length of DNA fragments has a linear relationship with the distance that the fragments move through an agarose gel. A plot of the fragment length of known standards against the mobility of those fragments can be used as a standard curve (known as a Ferguson plot; refer back to the experiment you did on protein purification where you also used standard curves) that allows the determination of the length of an unknown fragment by measuring its mobility.

Agarose gel electrophoresis can also be used to determine the quantity of DNA or RNA in a sample. DNA molecules are very good at absorbing ultraviolet light but not visible light. In order for us to see where they have moved in an agarose gel it is usually necessary to stain them with a dye (Figure 11.2). One commonly used dye, SYBR® Safe DNA gel stain, binds fairly specifically between the stacks of nucleotides in DNA double helices and, when in that position, can be caused to fluoresce when exposed to ultraviolet or blue light (which is first absorbed by the DNA molecule which then transfers its energy to the stain which in turn emits light visible to our eyes). The amount of light emitted by such stains is directly proportional to the amount of the stain that is bound to DNA. If DNA standards of known masses are present in the adjacent lanes of an agarose gel, molecular biologists can estimate the amount of DNA present in an unknown sample by comparing the intensity of its fluorescence to that of the known DNA fragments. This method is rapid, free from interference from contaminating molecules, and sufficiently accurate for most purposes.

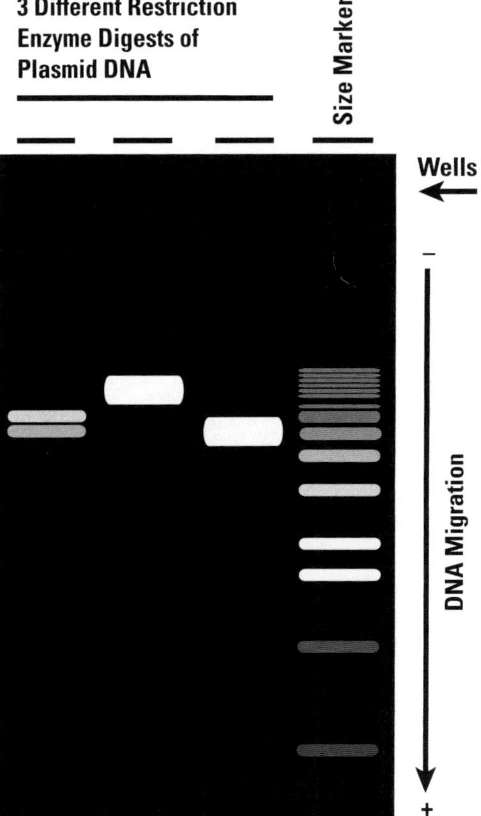

Figure 11.2 Digital image of an agarose gel stained with SYBR® Safe DNA gel stain. The first and third lanes of the gel contain fragments of a plasmid that has been cut with three different restriction enzymes. The fourth lane contains a size standard (fragments of DNA of known length).

PROCEDURES

In this course of the first step of this experiment, you will use agarose gel electrophoresis to determine the charge (positive or negative), relative mobility (R_f value) and identity of unknown dyes.

STEP ONE

Prior to your arrival in the laboratory, the prep-room personnel will have prepared several 1.5% agarose gels (1.5 grams of agarose per 100 mL of water based buffer) for you to use. The Tris/borate/EDTA buffer (TBE) in the gels will be at the same concentration as in the running buffer you will submerse the gels in during electrophoresis. TBE is a commonly used buffer in molecular biology laboratories and its three ingredients each serve an important purpose: Tris is a very effective buffer around neutral pHs, borate is a salt that carries the electric field you will apply, and EDTA seeks out and binds to bivalent cations like Mg^{2+} and Ca^{2+} which are essential to the activity of most enzymes that degrade proteins and DNA.

1. Work individually for this step of today's laboratory experiments. Take a micropipettor and one of the small microcentrifuge tubes numbered "1" through "40" from the equipment bench. Each of the microcentrifuge tubes contains a mixture of two dyes that differ in their charge and/or molecular weight. Record the number of your microcentrifuge tube and report it in the results section of your laboratory report.

2. Practice withdrawing and returning 10-μL quantities of the dye solution with your micropipettor. When you are comfortable with this procedure, move to the bench at the right side of the laboratory (the side closest to the prep room) and use the micropipettor to transfer 10 μL of your dye solution into one of the wells in the practice agarose gels.

3. After you have successfully loaded one of the wells in a practice gel, move to the front of the laboratory and ask your teaching assistant which of the submarine electrophoresis apparatuses are safe and ready for loading. Making sure that all the electrical leads to the apparatus are disconnected, use your micropipettor to load 10 μL of your dye into one of the unused wells of an agarose gel.

4. When all the wells of the agarose gel have been loaded by students, observe your teaching assistant connect the apparatus to a power supply and apply an electrical field to the agarose gel. Note which side of the gel is closest to the positive and negative electrodes of the apparatus.

5. After roughly 45 minutes of electrophoresis, the dyes in your unknown should be sufficiently separated for you to determine their identity. Your teaching assistant will disconnect the apparatus from the power supply and allow you to examine the way that the dyes in your solution have moved. Use a ruler to measure the distance between the center of the well and the center of the dye band for each of your dyes and record those values in the first row of Table 11.1. Positively charged dyes will have moved toward the negative electrode and the distances they have traveled should be assigned negative values.

6. Measure the distance that the amido-black your teaching assistant loaded on the same gel has moved and record that value in the first row of Table 11.1.

Table 11.1 Electrophoretic mobility of dyes in an agarose gel.

	First Unknown Dye	Second Unknown Dye	Amido-Black
Distance (cm)			
Color			Dark Blue
R_f Value			1.00

7. Record the colors of your separated dyes in the second row of Table 11.1.

8. The distance and direction that any molecule such as the dyes in your unknown move through a matrix during electrophoresis is dependent upon the molecule's charge and molecular weight. The distance traveled relative to a known standard should allow you to identify the two dyes in your unknown. In this case, the distance traveled by each of your dyes divided by the distance traveled by amido-black will be equal to a distinctive **retardation factor** (R_f). Calculate the R_f value for each of the dyes in your solution and record those values in the third row of Table 11.2.

9. Compare the colors and R_f values of the dyes listed in Table 11.2 to those in Table 11.1 to determine the identity of the dyes in your sample.

Table 11.2 Relative electrophoretic mobility of dyes in an agarose gel.

Dye Name	Color	R_f Value
Brilliant Cresyl Blue	Dark Blue	− 0.85
Safranin O	Light Red	− 0.71
Brilliant Green	Light Green	− 0.68
Gentian Violet	Purple	− 0.59
Bismarck Brown Y	Yellow	− 0.20
Xylene Cyanol	Light Blue	0.30
Rose Bengal	Pink/Red	0.72
Congo Red	Orange	0.79
Bromphenol Blue	Purple/Blue	0.90
Amido-black	Dark Blue	1.00
Orange G	Yellow/Orange	1.13

STEP TWO

As in Step One of this experiment, you will be using a 1.5% agarose gel (1.5 grams of agarose per 100 mL of water-based TBE buffer). Unlike the agarose gel you will have used in Step One, the buffer and agarose gel you will use for Step Two will also contain 0.5 µg/mL of SYBR® Safe DNA stain (ethidium bromide might be used as an alternative if this stain is not available). This stain will cause DNA molecules to fluoresce when exposed to ultraviolet (and deep blue) light. λ is a bacteriophage of *E. coli* that contains a single chromosome that is 48,502 nucleotides long. You will be using λ DNA that has been cut with the restriction enzyme *Hind*III as a standard (Figure 11.3) to determine the length and concentration of an unknown DNA molecule.

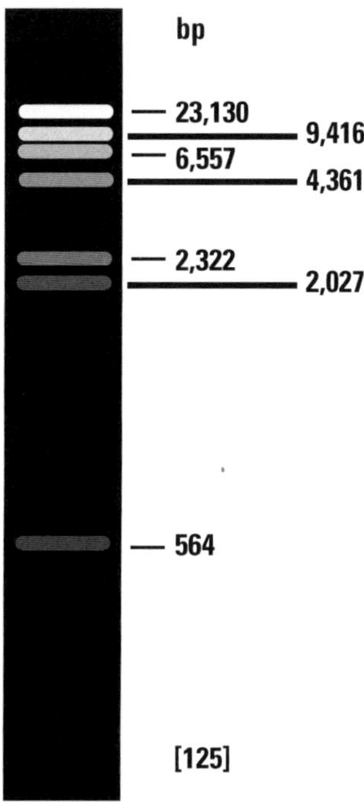

Figure 11.3 Bacterophage Lambda DNA × *Hind*III fragments when size fractionated on an agarose gel. The larger fragments move more slowly than the smaller ones from the top of the gel to the bottom. Larger fragments also bind more of the stain that allows us to see the DNA than smaller fragments. The size (in base pairs) of each fragment is shown to the right of the bands.

1. Take a micropipettor and one of the small microcentrifuge tubes labeled "U" from the equipment bench. Each of the microcentrifuge tubes contains a sample of a plasmid DNA that has been cut with the restriction enzyme *Hind*III to make it a linear DNA. Record the letter of your microcentrifuge tube and report it in the results section of your laboratory report.

2. Obtain a 1" strip of Parafilm® and use a micropipettor to add the following to three spots on the Parafilm® (prepare these samples for use close in time to when you will be loading them on an agarose gel—the water in them will evaporate). Spot #1: 9 µL water and 1 µL of loading dye. Spot #2: 20 µL of 0.05 µg/µL λ/*Hind*III DNA plus loading dye. Spot #3: 10 µL of your unknown DNA plus loading dye.

3. Set your micropipettor to deliver 10 µL each of spots 1 and 3, and 20 µL for spot 2. Proceed to a gel rig that contains a 1.5% agarose gel that has been prestained with SYBR® Safe DNA stain and load your three samples in order (1, 2, and 3) in three adjacent lanes. (The sample buffer contains two tracking dyes and glycerol that will help keep the samples in the bottom of the gel's wells.)

4. When all the wells in the gel have been loaded, your teaching assistant will apply an electric field to the gel. Periodically check on the status of the gel and ask your teaching assistant to stop it when the faster of the two tracking dyes is close to the end of the gel.

5. When the electrophoresis of the agarose gel is done, observe your teaching assistant as they take your gel to the transilluminator of the gel documentation station to capture a digital image of the gel.

6. Use a printout of your gel's image and a ruler to measure the distance that the leading edge of each DNA fragment in your second and third gel lane have moved and record those distances in Table 11.3.

7. For each of the fragments in the lane with the λ size standard, calculate the $\log_{10}$ of the length of each fragment and record those values in Table 11.3.

Table 11.3 Electrophoresis of standard and unknown DNA fragment in a 1.5% agarose gel.

λ DNA Fragment Size (bp)	$\log_{10}$ λ DNA Fragment Size	Distance λ DNA Fragment Moved (cm)	Distance Unknown DNA Fragment Moved (cm)
23,130			
9,416			
6,557			
4,361			
2,322			
2,027			
564			
125			

STEP THREE

In this step you will interpret the data you have collected in Step Two to determine the length and concentration of the unknown DNA fragment you electrophoresed. You can do this portion of the laboratory away from the lab, but you may find it helpful to complete it during your laboratory's meeting time where you will be able to get help from your teaching assistant or prep lab personnel in doing the analyses.

1. Use a piece of graph paper (provided in the laboratory) and the values in the second and third columns of Table 11.3 to construct a Ferguson plot (which you should include in your laboratory report). The distance the λ DNA fragments electrophoresed should be on your plot's horizontal axis and the $\log_{10}$ fragment size of each of the λ DNA fragments should be on the vertical axis of your plot.

2. Use a ruler to draw a line that best runs through the eight data points in your Ferguson plot. In drawing the line you will be performing a linear regression that describes the relationship between the size of a DNA fragment and the distance that it moved in the 1.5% agarose gel that you used. (Commonly available spreadsheets like Excel can automatically determine a best fit linear regression for data such as this and generate an equation that can then be used to determine the size of unknown DNA fragments on the basis of how quickly they move through an agarose gel.)

3. Use your Ferguson plot and the distance that your unknown DNA fragment electrophoresed to determine the $\log_{10}$ of the size (in base pairs) of your unknown DNA fragment. Be sure to include your estimate of your unknown DNA fragment's size in base pairs in your laboratory report.

4. Determine which λ DNA fragment in your λ standard gives rise to a band that is closest in intensity to the band from your unknown DNA fragment. You loaded 20 µL of 0.05 µg/µL λ DNA into the second well of the 1.5% agarose gel you used in Step Two which means that there is a total of 1.0 µg (or 1000 nanograms) of DNA in that lane.

5. Use the size of the fragments of l DNA provided in Table 11.3 to determine what fraction of that total amount of DNA in the lane corresponds to each band you can see (e.g., the largest band corresponds to 23,130/48,502 × 100% or 47.7% of the DNA in that lane which means that its band contains almost half of the 1.0 µg of DNA in that lane).

6. Knowing that the amount of fluorescence by SYBR® Safe DNA stain is directly related to the amount of DNA present in a band, how much DNA (in µgs) is present in the band from your unknown DNA fragment's sample? Remember that you loaded 10 µL of your unknown fragment in your third lane. What was the concentration (in µg/µL) of that DNA fragment in your unknown sample? Be sure to include your estimate in your laboratory report.

IMPORTANT QUESTIONS TO CONSIDER

Careful consideration of the following questions related to this set of experiments will not only insure that you understand the most important concepts of this laboratory but will also help you prepare for the lecture portion of the course. By addressing these questions (and others that you decide are important) in the conclusions section of your laboratory report, you will also assure that you receive full credit for that portion of your report.

- WHAT WERE THE two dyes you separated by electrophoresis?

- WHAT CAN YOU conclude about the charges and molecular weights of the dyes in your unknown relative to each other and to amido-black?

- HOW MIGHT THE R_f values for the dyes you electrophoresed be different if the gel and buffer you used were run at pH 11 instead of 7 (assume that amido-black's mobility would not be affected)? What if they were run at pH 3 (again, assume that amido-black's mobility would not be affected)?

- HOW WELL DID the $\log_{10}$ values for the size of the λ DNA fragments correlate with the distance that they traveled in your 1.5% agarose gel?

- WHAT WAS THE length in base pairs of your unknown DNA fragment? How confident are you in your estimate?

- WHAT WAS THE concentration of the unknown DNA fragment you were provided?

- HOW WOULD YOUR results have been different if the agarose gel you had used was 1.0% agarose? 2.0% agarose?

EXPERIMENT 12: DNA PROFILING

"The presence of a DNA profile says nothing about the time frame or the circumstances under which DNA came to be associated with an evidence sample." Dan Krane

OVERVIEW

In the previous experiments you have seen evidence that suggests that a large portion of the biological information passed from one generation to another is carried in DNA and you have studied the molecular mechanisms that transmit this "genetic blueprint." These same lines of inquiry have led to fundamentally important advances in such varied fields as evolution and human genetics. They have also led to the development of techniques such as forensic DNA profiling that have had a significant impact on society and the criminal justice system.

It would probably be difficult to find a person who was not aware of the pivotally important role that DNA profiling, essentially the characterization and matching of trace amounts of DNA, has played in criminal trials. Widely heralded as the most significant advance in forensic science since the advent of fingerprinting in the early 1900s, DNA profiling became an invaluable tool to our criminal justice system in less than ten years and is now widely regarded as being the gold standard of forensic sciences.

By simulating the interpretation of a DNA profile in this set of experiments, you will be exposed to one of many practical applications that are a direct result of the type of research you have been conducting in the laboratory. At the same time you will have an opportunity to assess the methodology's strengths and weaknesses and to evaluate the statistical implications of DNA profiles.

INTRODUCTION

DNA (deoxyribonucleic acid) is the **genetic material.** In other words, the information stored in DNA allows the organization of inanimate molecules into functioning, living cells. As a direct result, these organisms have the ability to regulate their internal chemical composition, growth, and reproduction as you have seen in the previous experiments you have performed in this laboratory. At the same time that it governs those complex processes, DNA is also the material that allows us to inherit our mother's curly hair, our father's blue eyes and winning smile, and even our uncle's too large nose. As you have seen, the units that govern those characteristics, be it chemical composition or nose size, are called genes.

Genes themselves contain their information as a specific **sequence** of nucleotides that are found in DNA molecules. Only four different nucleotides (sometimes also referred to as "bases") are used in DNA molecules: adenine, guanine, cytosine, and thymine (**A, G, C,** and **T**). All the information within each gene comes simply from the order in which those nucleotides are found. Complicated genes can be many thousands of nucleotides long. Many genes code for proteins and best estimates are that it takes approximately 250,000 different proteins to make a human being.

The sum total of an organism's genetic material is referred to as its genome. The human **genome** contains approximately 3.2 billion (3,200,000,000) nucleotides and virtually every cell that makes up our bodies (of which there are trillions) contains a nearly perfect copy of that genome. If the nucleotides of your own genome were typed out, they would fill dozens of complete sets of encyclopedias.

Errors are occasionally made during the copying of the long stretches of As, Gs, Cs, and Ts in DNA. When these errors occur in cells that give rise to eggs or sperm and are passed on to subsequent generations, a **mutation** is said to have occurred. Changes in the regions of DNA containing genes gives rise to all the variability we see between individuals but can also result in genetic diseases (such as sickle cell anemia, phenylketonuria, albinism, cystic fibrosis, and some forms of colon cancer). Since mutations in coding regions often do not allow an organism to pass on its genes (or even survive) as effectively as individuals that do not have them, regions that are important to an organism's survival are usually the same from one individual to another and sometimes between one kind of species and another. Such regions are said to be **conserved.**

For most of the genome, changes are accumulated slowly over many millions of years even when they occur in regions that are not conserved. It has been estimated that 98.5% of the DNA in gorillas is identical to that found in humans (Figure 12.1). Unrelated humans are about 99.5% similar to each other at the DNA level. Still, due to the huge size of the human genome, a 0.5% difference between individuals amounts to literally millions of nucleotide differences. If those differences could be detected reliably, it would be possible to uniquely identify virtually any diploid organism—including individual humans.

human HBβ: ATGGTGCACCTGACTCCTGAGGAGAAGTCTGCCGTTACTGCCCTGTGGGGCAAGGTG

human HBβ: ATGGTGCACCTGACTCCTG**t**GGAGAAGTCTGCCGTTACTGCCCTGTGGGGCAAGGTG

gorilla HBβ: ATGGT**a**CACCTGACTCCTGAGGAGAAGTCTGCCGTTACTGCCCTGTGGGGCAAG**a**TG

rabbit HBβ: ATGGTGCA**t**CTG**t**C**cag**TGAGGAGAAGTCTGC**g**GT**c**ACTGCCCTGTGGGGCAAGGTG

chicken HBβ: ATGGTGCAC**tg**GACT**g**CTGAGGAGAAG**cagct**Ca**T**cAC**c**G**g**CCT**c**TGGGGCAAGGT**c**

Figure 12.1 Multiple alignment of nucleotide sequences from the β-like globin gene cluster. The sequences shown correspond to the first 57 nucleotides of the gene that codes for the β globin (HBβ) protein that is used to make hemoglobin. Lower case letters indicate differences in nucleotide sequence relative to the human sequence in the first line. HBβ is a naturally occurring variant of HBβ—individuals with this variant suffer from sickle cell anemia due to this single difference in the nucleotide sequence of their DNA.

Obtaining the sequence of an individual's entire human genome, however, is still a daunting task under the very best of circumstances. Fortunately, though, some relatively small regions of the human genome have been found which are not only not conserved but are actually **hyper-variable.** Many of the millions of differences between individuals appear to be concentrated in these regions. Some loci have such high levels of variability that it is often necessary to look at ten to 100 individuals before any two are found to be the same.

A group of English scientists led by Alec J. Jeffreys discovered hypervariable loci like these in 1985. The subsequent examination of these loci is now commonly referred to as **DNA profiling.** The loci that are commonly used for forensic and paternity testing purposes today are regions that contain variable numbers of short tandem repeats, or STRs. The "tandem repeats" at these loci are actually end-to-end duplications of short stretches of nucleotides (Figure 12.2). They are especially useful to forensic scientists because the difference in the DNA sequences between alleles can be easily detected from very small amounts of starting material simply by determining the length of the locus (e.g., the number of times the repeat sequence is repeated) without ever even having to "read" the sequence of As, Gs, Cs, and Ts (again, a labor and time intensive process).

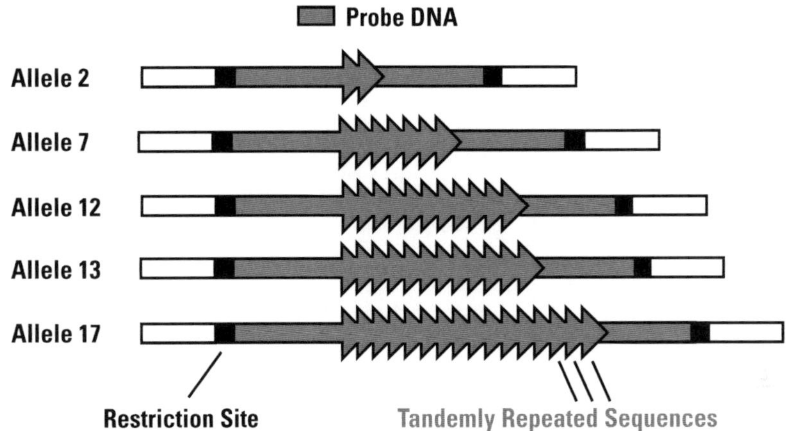

Figure 12.2 Schematic of a series of VNTR alleles. Five alleles are shown to differ in length due to varying numbers of tandem repeats (arrow heads). Filled boxes represent the primer binding sites at which PCR amplification of the fragments containing the tandem repeats would begin and end. The relative lengths of these fragments could then be determined by electrophoresis.

The sensitivity and discriminating power of DNA profiling both took quantum leaps with the integration of **polymerase chain amplification (PCR)** in the mid 1990s. Molecular biologists routinely use PCR as a sort of molecular copying machine that can make billions of copies of portions of DNA molecules that are of particular interest (like those that are likely to differ from one individual to another). Short stretches of complementary DNA known as primers guide DNA polymerases to the region to be amplified (Figure 12.2). Pairs of primers not only define the beginning and end of the region to be amplified, they also bring fluorescent dyes that label the amplified DNA fragments in a way that makes them easy to detect. DNA profiles used in criminal investigations today are routinely generated from commercially available kits that use PCR to amplify 13 highly **variable short tandem repeat (STR)** loci. Capillary electrophoresis is used to separate PCR amplified fragments on the basis of their size and results are documented in **electropherograms** (Figure 12.3). Alleles at these STR loci are named in a way that reflects the number of times that four nucleotide long sequences are tandemly repeated (i.e., an allele with 15 copies of the tandem repeat is referred to as a "15 allele" and one with 17 copies is referred to as a "17 allele").

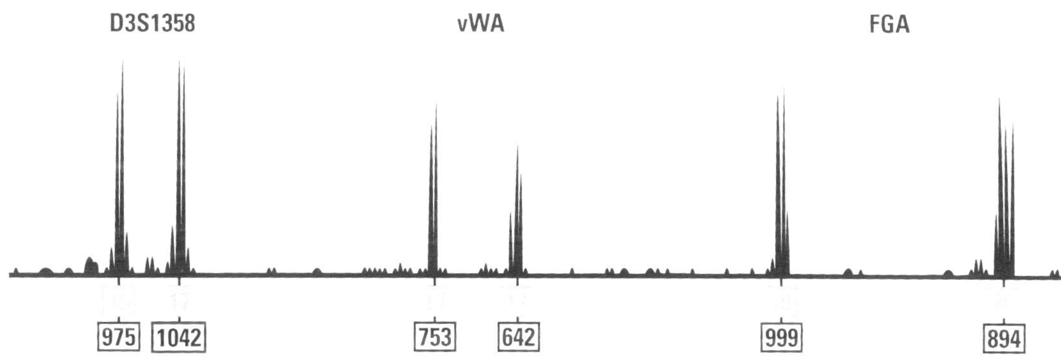

Figure 12.3 Electropherogram showing the DNA profile of an individual at three STR loci (D3S1358, vWA, and FGA, from left to right). Larger DNA fragments are to the right. Peaks correspond to alleles detected and are labeled with the name of the allele. The heights of peaks (shown in boxes beneath the allele labels) are correlated with the amount of DNA present in a sample prior to PCR amplification.

The power of the whole methodology of DNA profiling (both past and present) comes from the fact that a great deal of information can be obtained from a very small amount of starting material. STR DNA profiles can be generated from just a few cells. We lose dozens of cells without thinking about them every time we touch an object. DNA profiling results that exclude suspects are incontrovertible (barring laboratory error). If it is found that an individual has STR alleles of a certain size not found in evidence at a crime scene, it can be said with great confidence that that individual did not contribute that biological material.

Inclusion of suspects whose alleles do match those found at crime scenes however can never be established with absolute certainty. Instead, forensic scientists usually report the probability that a randomly chosen, unrelated individual would have the same alleles.

If a pair of alleles from a single locus are found in only one individual in 100, a match between the alleles seen in a suspect's genome and those found in evidence left at a crime scene could be convincing in itself.

What is the probability that a randomly chosen, unrelated individual would have the same set of alleles across 13 STR loci as those found in both a suspect and in a biological sample from a crime scene? Typically, forensic labs answer that question by invoking a statistical principle known as the **product rule.** With it, the probability of several unlikely events (i.e., having a particular STR allele at a locus) occurring simultaneously (i.e., having a particular combination of 26 STR alleles) is determined by multiplying together the probabilities of each of those events happening separately. Thus, if the chance that a randomly chosen person would have one particular STR allele at a locus is 1/10 and the chance that they would have another particular allele at that locus is 1/5, the chance of someone else having that particular pattern of pair of alleles can be calculated with $2pq$ (from Equation (10.1)) and would be said to be $2 \times 1/10 \times 1/5$, or 1/25. The more loci examined the more dramatic the numbers in the end. Odds as impressive as one in a quadrillion (one thousand million million!) are routinely obtained in this way.

The chances of finding two randomly chosen individuals whose DNA profiles are the same at 13 or more loci are slim. But evidence samples are often prone to a large variety of alternative interpretations (due to mixtures and/or the loss of information) that result in dramatic reduction of the statistical weight that can be attached to a "match."

PROCEDURES

Crime laboratories typically use a set of allele frequencies that were determined from the genotyping of a relatively small number (less than 200) of individuals (Table 12.1).

Table 12.1 Allele frequencies within the North American Caucasian population for three commonly used STR loci. These frequencies are derived from Hill et al. (Forensic Science International: Genetics; 2013, 7: e82-e83).

Allele	D3S1358	vWA	FGA
10.0	0.0014		
11.0	0.0014		
12.0		0.0014	
13.0	0.0014	0.0014	
14.0	0.1066	0.0928	
15.0	0.2729	0.1053	
16.0	0.2382	0.2008	
17.0	0.2105	0.2839	
18.0	0.1510	0.2022	0.0249
19.0	0.0166	0.1039	0.0499
20.0	0.0014	0.0069	0.1233
21.0		0.0014	0.1787
21.2			0.0055
22.0			0.2050
22.2			0.0125
23.0			0.1524
23.2			0.0028
24.0			0.1343
24.2			0.0014
25.0			0.0789
26.0			0.0263
27.0			0.0042

Table 12.2 Allele frequency values for unrelated Caucasians U.S.A. population samples. These frequencies are derived from Hill et al. (Forensic Science International: Genetics; 2013, 7: e82-e83).

Allele	D2S441	D21S11	D7S820
7.0			0.0277
8.0			0.1440
8.1			0.0014
9.0	0.0014		0.1676
9.1	0.0014		
10.0	0.2105		0.2562
11.0	0.3435		0.2050
11.3	0.0609		
12.0	0.0471		0.1593
12.3	0.0042		
13.0	0.0291		0.0346
14.0	0.2410		0.0042
15.0	0.0596		
16.0	0.0014		
25.2		0.0014	
27.0		0.0222	
28.0		0.1593	
29.0		0.2022	
29.2		0.0028	
30.0		0.2825	
30.2		0.0291	
31.0		0.0720	
31.2		0.0983	
32.0		0.0055	
32.2		0.0900	
33.0		0.0014	
33.2		0.0263	
34.2		0.0042	
35.0		0.0014	
36.0		0.0014	

STEP ONE

1. At the beginning of the laboratory section, your teaching assistant will give you a simulated DNA profile. Make note of the case number or suspect number that is written on the printout and include that information in the results section of your laboratory report.

2. If your printout has a case number and two electropherograms with peaks in them, you are a "victim" in this exercise. If your printout has a suspect number and only one electropherogram with peaks in it, then you will play the role of a "suspect" in this exercise. If you are a "suspect," proceed to point **5** of this exercise and if you are a "victim" continue with point **3**.

3. As a victim, your DNA profile for a set of three loci will be in the top electropherogram. The other electropherogram (labeled "Evidence") will represent the DNA profile of biological materials collected at a scene of a crime. It is not uncommon for such evidence to contain a mixture of both the perpetrator's and victim's genetic material. Try to use the heights of the peaks to conceptually "subtract out" your contributions to the evidence sample to learn the alleles contributed by a perpetrator.

4. Use the allele frequencies provided in Table 12.1 and the Hardy-Weinberg Equation (10.1) to determine the frequency that you would expect to find the combination of alleles in someone other than the perpetrator.

   For example, if a DNA profile has peaks labeled 15 and 17 at the D3S1358 locus (see Figure 12.3), the frequency of individuals in the population with matching DNA profiles is equal to $2pq = 2 \times 0.246 \times 0.212 = 0.0522$. It is often easier to interpret these frequencies if they are presented as probabilities (the reciprocal of the frequency). In this example the frequency would be 1/0.0522 or 1 in 19.2 meaning that you would expect to need to examine the DNA profiles of approximately 19 individuals to find one other that is indistinguishable from that of the perpetrator of the crime in your case. Perform this calculation separately for each of the three tested loci. Then multiply those three frequencies together to obtain an overall genotype frequency. Be certain to include these calculations in the results section of your laboratory report.

   Proceed to point **6** of this exercise.

5. As a suspect, use the allele frequencies provided in Table 12.1 and the Hardy-Weinberg Equation (10.1) to determine the frequency that you would expect to find this combination of alleles if these three loci are in Hardy-Weinberg equilibrium.

   For example, if a DNA profile has peaks labeled 15 and 17 at the D3S1358 locus (see Figure 12.3), the frequency of individuals in the population with matching DNA profiles is equal to $2pq = 2 \times 0.246 \times 0.212 = 0.0522$. It is often easier to interpret these frequencies if they are presented as probabilities (the reciprocal of the frequency). In this example the frequency would be 1/0.0522 or 1 in 19.2 meaning that you would expect every nineteenth DNA profile examined at this locus to be indistinguishable from that in your reference sample. Perform this calculation separately for each of the three tested loci. Then multiply those three frequencies together to obtain an overall genotype frequency. Be certain to include these calculations in the results section of your laboratory report.

6. When your teaching assistant has determined that everyone is ready, the suspects will be asked to form a line along one of the side walls of the laboratory and victims will be asked to form another line along the front of the laboratory. The line of victims will then pass by the suspects and examine their DNA profiles until a match to the profile in their evidence lane is found.

7. Once victims have found a matching DNA profile, they should return to their laboratory bench. Suspects, however, should remain standing until the last victim is seated (each suspect may be involved in multiple cases or none at all).

8. When everyone has returned to their seats, take five minutes to consider the implications of your DNA profile results and the arguments that prosecuting and defense attorneys might use in presenting this evidence to a jury. For instance, assume that the frequencies provided in Table 12.1 were based on a sample of only 200 individuals, all with a racial background different from the suspect and/or victim in the case—what affect will those common problems (small sample sizes and un-representative sampling) have on the significance of a DNA profile match? Most testing laboratories score false positives at a rate of once every one hundred to five hundred tests that they perform—might that alter the weight attached to the matches you observed? Were mixtures resolved accurately? Could there have been more than two contributors to the mixed samples?

9. Once everyone has had an opportunity to consider their "cases," your teaching assistant will ask for volunteers to present their evidence in a mock trial. If both the "victim" and "suspect" in your case are willing to play the role of "prosecutor" and "defense attorney," respectively, your teaching assistant will take on the role of judge and your lab mates the role of jurors. Winners of their cases, as determined by the jury of your peers, will automatically receive a full ten points for this laboratory exercise and will not be required to turn in answers to the questions at the end of this experiment.

IMPORTANT QUESTIONS TO CONSIDER

Careful consideration of the following questions related to this set of experiments will not only insure that you understand the most important concepts of this laboratory but will also help you prepare for the lecture portion of the course. By addressing these questions (and others that you decide are important) in the conclusions section of your laboratory report, you will also assure that you receive full credit for that portion of your report.

- DID YOU SEE any evidence of contamination and/or degradation of DNA in the DNA profiles you examined?

- HOW IMPORTANT IS the database that is used to determine allele frequencies in DNA profiling cases?

- OUR COUNTRY IS often referred to as a "melting pot" but that term is probably more appropriately applied to the manner in which we quickly assimilate each other's cultures rather than our genes. Blonde hair and blue eyes, for instance, are still predominantly found in individuals of German or Scandinavian ancestry even though many of the cultural attributes of those groups (such as an appreciation of beer) are found throughout our society. How might you take into account population substructuring (ethnic groups or other collections of individuals that prevent the totally random exchange of genetic material) in determining the significance of a DNA profile match?

- WHAT FACTORS OTHER than population substructuring might make a database unreliable and how would you minimize their effect?

- HOW WOULD YOU determine if two alleles or two sets of genes such as those that code for blonde hair and blue eyes are correlated or linked in a population?

- WHAT EFFECT WOULD looking at more than just three loci have on the frequencies and probabilities you calculated for your case?

- WHAT STEPS COULD forensic scientists take to minimize the chances of false positives or contamination in their work?

- HOW HERITABLE ARE DNA profiles relative to fingerprints?

- WHY SHOULD DNA analysts draw conclusions about what alleles are associated with an evidence sample before they know a suspect's DNA profile?

You can learn a great deal more about forensic DNA profiling and about the research being done here at Wright State that is making the current STR-based methods of DNA profiling more objective at the following web site: *www.bioforensics.com.*

EXPERIMENT 13:
BACTERIAL TRANSFORMATION

OVERVIEW

Biotechnology has become an enormous industry over the past forty years by harnessing the ability of living systems to quickly and inexpensively produce commercially important biological materials. Almost every aspect of our day-to-day lives is affected by biotechnology from the food we eat to the drugs prescribed by our physicians. At the very heart of this industry is our ability to permanently alter an organism's set of genetic instructions in a process known as transformation.

Over the past several weeks you have studied the way in which cells regulate their chemical composition, growth and division and how these processes can change over the course of time as allele frequencies are altered. In simulating DNA profiling, you have also seen one of the many ways in which an understanding of naturally occurring variation in alleles can have a very practical application.

In this set of experiments, you will use your understanding of those natural processes to by-pass nature's random fashion of generating variation. Through transformation you will genetically engineer a new organism with its own set of distinctive characteristics. This is a heady prospect—the resulting organism will exist solely because you chose for it to exist and applied your skills to the task of creating it. In the course of these experiments, you will also have an opportunity to use gel-electrophoresis to separate molecules on the basis of their charge and size—another important tool of biotechnology and molecular biology in general.

INTRODUCTION

Whether or not the variability observed in bacteria was due to true genetic differences or simply to adaptive responses was a matter of considerable debate among microbiologists for most of the first half of this century. In 1943, however, S. E. Luria and M. Delbrook settled the issue when they proved that bacteria, just like pea plants and fruit flies, underwent spontaneous mutations that altered their genetic material. In answering that single question, Luria and Delbrook also initiated a field of study that has revolutionized the study of biology—microbial genetics.

Bacteria have proven to be ideal organisms for genetic manipulation and have been extensively used in recombinant DNA research. Their short division time, relative simplicity, and haploid nature allow the effects of manipulations of their genetic material to be readily assessed.

One common bacteria, *Escherichia coli (E. coli)*, has been particularly well-characterized and is widely used in molecular biology laboratories. A natural resident of the human gut, *E. coli's* genome is approximately three million nucleotides long (about one thousandth the size of the human genome). Every gene in this organism has been identified, and all four million nucleotides of its genome are available on the internet as part of the *E. coli* genome sequencing project. Most of *E. coli's* genetic information is carried on its single chromosome, but bacterium often contain one or more small pieces of extrachomosomal DNA as well.

These smaller DNA molecules, **plasmids,** range in size from 1,000 to 100,000 nucleotides in length and can also carry genetic information that is vital to the cell. In nature, plasmids commonly carry genes that allow bacteria to neutralize antibiotics that would otherwise kill them. These antibiotic resistance genes provide scientists with a very convenient means of **selecting** bacteria that have acquired a particular plasmid—quite simply, if a bacteria can live in the presence of an antibiotic like ampicillin or tetracycline, then it is very likely to harbor a particular plasmid.

Plasmids also provide a convenient vehicle for the acquisition of novel genetic information by bacteria. For instance, growing concern about the increasing resiliency to antibiotics of the bacteria that cause tuberculosis is largely due to plasmids that these bacteria have recently acquired. Whenever a transfer of genetic information between organisms takes place *via* the movement of an extracellular piece of DNA, a **transformation** is said to have occurred.

Transformation probably occurs most often in nature when bacterium come upon the plasmids of another cell that has been lysed (broken open) by an invading virus. The precise mechanism that causes bacteria to transport these relatively large molecules across their membranes remains a mystery. Nonetheless, scientists capitalizing on this phenomenon have been rewarded with several important discoveries. For instance, recall that Griffith's work with the bacterium *Streptococcus pneumoniae* in 1928 was the first step toward demonstrating that DNA was the genetic material. Subsequent experiments by O. T. Avery and C. M. MacLeod using the same system revealed that DNA was the only biological macromolecule that was capable of transforming a noninfectious bacterium into a virulent strain that could kill mice.

Despite this long history of fruitful utilization by scientists, most methods for bacterial transformation today are based on an observation of M. Mandel and A. Higa. In 1970, these virologists discovered that bacterial uptake of naked viral DNA was enhanced by simultaneously exposing bacterial cells to calcium chloride ($CaCl_2$). Many variations in this basic technique have since been described, all directed toward increasing the efficiency with which cells are made **competent** (able to take up foreign DNA). Most protocols currently yield 10^5 to 10^7 (100,000 to ten million) transformed bacteria per *micro*gram (1/1,000[th] of a gram) of intact plasmid DNA. As impressive as those numbers are, it is important to bear in mind that under these optimized conditions not more than one DNA molecule in 10,000 finds its way into a competent cell and is successful at transforming it.

From a biotechnology and evolutionary perspective however, the creation of a single transformant is quite sufficient. Once a cell has been transformed, its genetic information is permanently supplemented and it faithfully passes on its new characteristics to all its progeny. If for example, a biotechnology company constructed a plasmid that carried not only a gene that conferred resistance to ampicillin but also a gene that produced β globin, large quantities of that important protein could be purified from transformed bacteria. As long as the transformed bacteria are able to survive in an environment that contained ampicillin they will continue to produce a form of β globin that is indistinguishable from the β globin produced by humans.

The construction of such genetically engineered plasmids through the use of **recombinant DNA technology** has become quite routine in both research and industrial settings. Restriction enzymes have proven to be very useful in this work as well as in the application of DNA profiling described in the previous experiment of this laboratory manual. Many restriction enzymes such as *Eco*RI make staggered cuts at their specific restriction sites in double stranded DNA (Figure 13.1). These staggered breaks are commonly referred to as **sticky ends** due to the fact that their short stretches of single-stranded sequence are complementary and are capable of base pairing with each other. As a result, DNA from two or more sources that have been cut with the same restriction enzyme will have sticky ends that facilitate their being joined into a new DNA molecule.

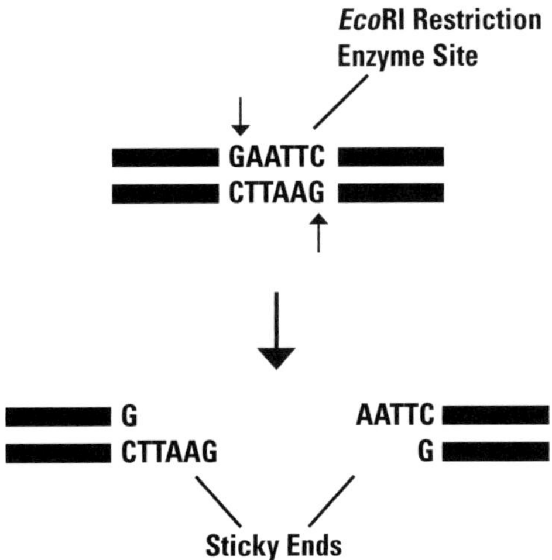

Figure 13.1 Molecular view of a restriction enzyme digestion. Many restriction enzymes break phosphodiester bonds within their restriction sites in a way that generates sticky ends. The cleavage sites for *Eco*RI and the sequences on the end of the fragments it generates are shown.

New plasmids can be very specifically constructed through the careful selection of restriction enzymes as well as selection of the resulting DNA fragments that are joined together. As you learned in the set of exercises you performed last week, DNA fragments can be easily separated by their size through electrophoresis on agarose gels.

In general terms, electrophoresis relies upon the ability of electric fields to move charged particles as described in the materials associated with "Experiment 10: Human and Population Genetics." Most molecular biology laboratories separate protein and DNA molecules of different sizes and charges by utilizing matrices made from either polyacrylamide or agarose gels much like the one that you used in "Experiment 10: Human and Population Genetics."

PROCEDURES

As you learned in the introductory material, bacteria such as *E. coli* are usually unable to survive in the presence of ampicillin. In the following experiment you will transform competent *E. coli* cells with a plasmid carrying a gene that confers resistance to that antibiotic.

Only the cells that are successfully transformed will be able to survive and multiply into easily observed colonies on agar plates containing ampicillin. It will be necessary to allow the transformed cells to grow on the agar plates for at least eight (8) hours before these colonies are visible and you can see the final results of this experiment. To complete your laboratory report for this set of experiments, both you and your laboratory partner will need to make plans to **re-visit the laboratory at some point in the next week other than your regularly scheduled section to examine these plates**.

The *E. coli* bacteria you will be working with in the first step of this experiment are natural inhabitants of your intestines and pose no serious threat of infection if minimal precautions are taken in their handling. Still, it is important to practice "good sterile technique" when dealing with any micro-organism if for no other reason than to avoid cross-contamination of solutions. At the beginning of your laboratory section, your teaching assistant or the prep room personnel will demonstrate the basics of sterile technique that you will need to follow in the first step of this experiment.

STEP ONE

1. Work in groups of four. Obtain a beaker and two microcentrifuge tubes from the equipment bench. Use a pen to mark one of the microcentrifuge tubes "+" and the other "−". Fill the beaker with ice and insert the microcentrifuge tubes into the surface of the ice.

2. Using the sterile methods demonstrated by your teaching assistant or the prep room personnel, use a pipette to add 150 µL of cold 0.1 M $CaCl_2$ to two microcentrifuge tubes and return them to the ice.

3. Obtain a plate containing *E. coli* bacterial colonies from the equipment bench. Sterilize an inoculating loop by placing it in the flame of a Bunsen burner until it glows red. Cool the inoculating loop by poking it into the agar on the plate containing *E. coli* in a region that does not contain any bacterial colonies. Use the cooled, sterile inoculating loop to transfer one of the *E. coli* colonies into the tubes containing 150 µL of cold $CaCl_2$. Mix the contents of the tubes by tapping them with your finger or using a vortexer.

4. Transfer 10 µL of the solution containing plasmid DNA (0.005 µg/µL) to the cell suspension in the microfuge tube labeled "+" and mix the contents of the tube by tapping it with your finger. Make note of the time and return the microcentrifuge tube to the ice. The plasmid you are using in this experiment contains a gene that codes for β-lactamase. β-lactamase is an enzyme that inactivates the antibiotic ampicillin and confers resistance to that drug upon cells that can produce it.

5. In the next fifteen minutes, obtain two "LB" agar plates and two "LB+amp" agar plates from the equipment bench (LB is an abbreviation for Luria broth—a rich media for bacterial growth; amp is a common abbreviation for the antibiotic ampicillin). Use a pen to label one of the LB plates "LB+" and the other "LB−". Label one of the LB+amp plates "LB/amp+" and the other "LB/amp−". Label the lids of all four plates with your initials.

6. After 15 minutes have elapsed since you completed portion 4 of this experiment, heat shock the cells in your microcentrifuge tubes by placing them in a 42°C water bath for 30 seconds. Return the cells to ice at the end of this heat shock. This brief exposure to heat induces the expression of genes within the bacteria that help it recover from damage inflicted by exposure to the calcium chloride solution and greatly increases the efficiency of transformation.

7. Use a pipette to add 250 µL of liquid LB to each of your microfuge tubes. Seal the tubes and mix their contents by tapping with your finger. Set the tubes down on your laboratory bench and return the beaker containing ice to the equipment bench.

8. Use sterile technique to transfer 100 µL of the cells in the microfuge tube labeled "+" to the center of the "LB+" and "LB/amp+" agar plates.

9. Use sterile technique to transfer 100 µL of the cells in the microfuge tube labeled "−" to the center of the "LB−" and "LB/amp−" agar plates.

10. Spread the cells on all four plates using a **sterile** spreading rod. Sterilize the spreading rod before spreading the bacteria on each of the four plates to prevent cross-contamination. The spreading rod is sterilized by removing it from its ethanol bath, touching it to a Bunsen burner's flame and waiting for the ethanol to burn completely. After each use, return the spreading rod to its ethanol bath.

11. Use a single piece of tape to hold all four of your plates together, turn them upside down, and give them to your teaching assistant for incubation overnight at 37°C. Be certain to find out when you will be able to view the incubated plates **before** your next laboratory section so you can include the results from this experiment in your laboratory report.

12. After your plates have incubated overnight, examine them for bacterial growth. Count the number of distinguishable colonies on each of your four agar plates. If bacterial colonies are so abundant that they cover the entire surface of your plate, estimate the count to be "> 10,000." Record those counts in Table 13.1 and be certain to include them in the results section of your laboratory report. Each colony you observe on the "LB/amp" agar plates corresponds to a single cell in the "+" or "−" microcentrifuge tubes that was resistant to ampicillin.

Table 13.1 Bacterial colonies resistant to the antibiotic-ampicillin.

	LB−	LB+	LB/amp−	LB/amp+
Colonies				

IMPORTANT QUESTIONS TO CONSIDER

Careful consideration of the following questions related to this set of experiments will not only insure that you understand the most important concepts of this laboratory but will also help you prepare for the lecture portion of the course. By addressing these questions (and others that you decide are important) in the conclusions section of your laboratory report, you will also assure that you receive full credit for that portion of your report.

- GIVEN THAT ONLY one DNA molecule in 10,000 can successfully transform a bacterium in the most favorable conditions we can create in the laboratory, how important a phenomenon is transformation of bacteria in nature?

- WHY WOULD YOU not expect to see evidence of bacterial growth on the agar plate labeled "LB/amp–"?

- WHY WOULD YOU expect to see a difference in the number of bacterial colonies on the agar plates labeled "LB+" and "LB/amp+"? What does that tell you about the efficiency of the transformation you performed?

- IN WHAT MEDIA would you grow your transformed cells if you wanted to be certain that none of the bacteria lost their copy of the plasmid with the β-lactamase gene?

EXPERIMENT 14: DNA ISOLATION

OVERVIEW

Much of the work you have done in the laboratory over the past thirteen weeks has illustrated the central role that DNA plays in living organisms. Your experiments have illustrated that, as the repository of the bulk of an organism's genetic information, DNA dictates how it will grow, what energy sources it will use, how likely it is to pass its genes on to subsequent generations and how distinguishable it is from other living things.

Its many and varied roles may actually cause "DNA" to remain a relatively abstract concept even after all your experimentation. Most students find that they have a whole new appreciation for our genetic material when they isolate it from living cells and hold the literally billions of genes within it in their hands.

Despite its unique importance to cells, DNA has a set of chemical characteristics that allow it to be purified and concentrated much like any other biomolecule. In this experiment you will take advantage of the subtle chemical differences between DNA and other molecules. These differences will allow you to extract and purify it from onion roots and precipitate it in a form that you will be able to keep as a souvenir of BIO 1120 for decades.

INTRODUCTION

In the 1860s F. Miescher assigned the name "nuclein" to an acidic substance he isolated from the nuclei of cells. As his work progressed, he eventually coined the name for this material that we know it by today—nucleic acids. An additional 80 years were required before Griffith, Avery, and MacLeod were able to demonstrate the fundamental importance of this material and it became clear that it was in fact *the* genetic material.

Two general classes of nucleic acids can be found in every living organism; ribonucleic acids (RNA) and deoxyribonucleic acids (DNA). As implied in their names, the primary structural difference between the RNA and DNA is in the number of hydroxyl groups attached to their sugars. Both RNA and DNA are long polymers of **nucleotides** each of which are made from three chemical components: an inorganic **phosphate,** a monosaccharide **sugar** and either a **purine** or a **pyrimidine** nitrogenous base.

Each phosphate group within a nucleotide monomer is attached to its sugar by way of an ester bond. Nucleotide monomers are strung together into polynucleotides by the formation of an additional ester bond between the phosphate group of one monomer and the sugar of another. The resulting covalent linkages are often referred to as phosphodiester bonds and constitute the backbone of polynucleotide chains. The nitrogenous base of each nucleotide is also attached to the sugar but plays no role in covalently holding the polymer together. Rather, it is the ability to arrange purines and pyrimidines in any order along the invariant sugar-phosphate backbone that allows nucleic acids to act as information repositories.

The size range of polynucleotides is quite large. At one end of the spectrum are small nuclear RNAs (snRNAs) and transfer RNAs (tRNAs) that are strings of from 70 to 210 nucleotides while at the other end are eukaryotic chromosomes that can be tens of millions of nucleotides long and actually rank among the largest single molecules known.

The intrinsic chemistry of nucleic acids allows them to be separated from the other molecules present within cells and viruses. The properties listed in Table 14.1 have proven to be particularly helpful to molecular biologists interested in isolating and studying these materials. As you have seen in the previous two sets of experiments, nucleic acids themselves can also be effectively separated from each other on the basis of their relative sizes by gel electrophoresis. **Density gradient centrifugation** also provides a means of separating nucleic acids on the basis of their size as well as their shapes and is routinely used in laboratories to distinguish between circular and linear forms of the same DNA molecules.

Table 14.1 Physical and chemical properties of RNA and DNA.

DNA Properties	RNA Properties
Insoluble in Dilute NaC	Soluble in Dilute and Concentrated NaCl
Soluble in Concentrated NaCl	Generally Small Molecular Weights
Insoluble in Alcohols	Insoluble in Alcohols
Unaffected by Detergents	Unaffected by Detergents
Impervious to RNases	Impervious to DNases
Impervious to Proteases	Impervious to Proteases

Of course, both DNA and RNA are capable of adopting secondary structures due to hydrogen bonding (base pairing) that can occur between complementary sequences of nitrogenous bases. And, like the proteins you studied in "Experiment 3: Osmosis and Diffusion," most nucleic acids exist in ordered conformations with very specific secondary structures in cells. DNA in particular is well known for its strong tendency to form a double stranded helical structure that resembles a twisted ladder with each rung corresponding to the pairing between two complementary nucleotides.

As you might expect, certain treatments like exposure to high temperatures can disrupt the ordered secondary structures of nucleic acids. As temperature increases, the hydrogen bonding between base pairs is increasingly disrupted and the nitrogenous bases become more exposed. Consequently, the **melting** of DNA that occurs during thermal denaturation is also accompanied by a **hyperchromic effect**—an increase in the molecules' ability to absorb ultraviolet light due to the absorbance of the purines and pyrimidines in those wave lengths. The melting temperature (T_m) at which two DNA strands separate can thus be monitored and used to estimate the relative abundance of Gs and Cs to As and Ts within a given DNA molecule—the greater the number Gs and Cs (G + C content), the more hydrogen bonds will hold the complementary strands together and the greater the T_m.

Naturally, the most accurate means of determining the G + C content of any DNA molecule comes from an analysis of the nucleotide sequence itself. A. Maxam and W. Gilbert won a Nobel prize in 1980 for a sequencing method that relies heavily upon chemical degradation of polynucleotides in a nitrogenous base-specific manner. While the Maxam-Gilbert method is still used for special applications, most DNA sequencing in contemporary molecular biology laboratories is based on a safer and more sensitive approach devised by F. Sanger in the early 1980s.

Sanger's method differs primarily from Maxam and Gilbert's in its use of enzymatic rather than chemical techniques. In it, specially modified nucleotides are incorporated into growing polynucleotide chains by the same enzyme cells use to replicate DNA during cell division. The modified nucleotides differ from those that the enzyme normally encounters in that a hydrogen is substituted in the place of one of the hydroxyl groups on its deoxyribose sugar—they are **dideoxy nucleotides** (ddNTPs). When these modified nucleotides are incorporated into a growing DNA molecule, the DNA replicating enzyme is not able to make a phosphodiester bond between it and the next nucleotide that would otherwise appear in the polynucleotide chain.

In practice then, a small amount of a specific dideoxy nucleotide was provided to the enzyme along with each of the four nucleotides it normally uses. The resulting products were a series of chains that were specifically terminated at positions where the <u>di</u>deoxy nucleotide had been incorporated. By separately conducting these sequencing reactions with dideoxy forms of each of the four nucleotides (ddGTP, ddATP, ddTTP, and ddCTP), it was possible to generate a set of DNA molecules that can be separated by size into a "sequencing ladder" on a polyacrylamide gel (Figure 14.1). Each rung on the ladder corresponds to a single nucleotide in the DNA molecule being sequenced. Sequencing is typically accomplished today in a single reaction by having each of the four different dideoxy nucleotides labeled with a different fluorescent dye. It can be determined which dideoxy nucleotide led to a chain termination event simply by determining what color it fluoresces.

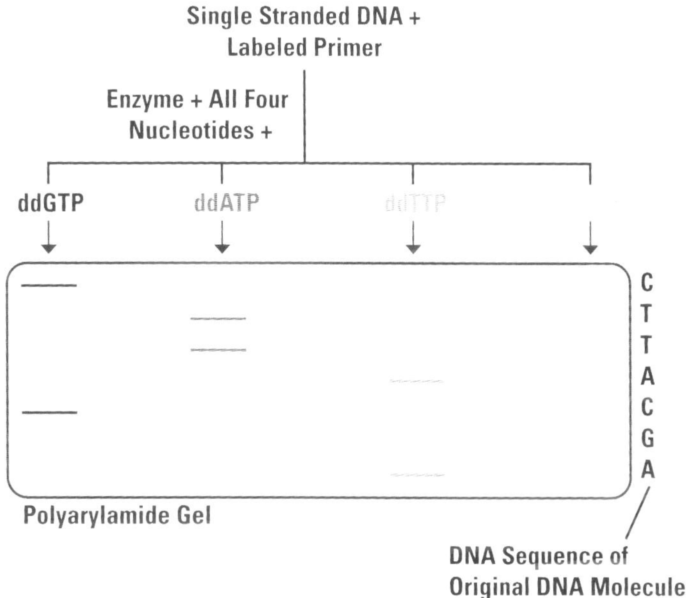

Figure 14.1 The Sanger dideoxy-DNA sequencing method. A strand of DNA to be sequenced is mixed with a labeled primer and split into four reaction tubes. Each tube contains enzyme, all four deoxynucleotides and a small amount of one of the dideoxy-nucleotides. The fragments that result can be separated by size on a polyacrylamide gel and their patterns can be used to determine the order of nucleotides in the original DNA strand.

And, of course, it is worth remembering that, before the information content of any DNA molecule can be determined in this way, the molecule itself must first be isolated. That important step is accomplished in research laboratories by employing techniques much like those you will use to purify banana DNA in this experiment.

PROCEDURES

The initial step in any DNA isolation involves breaking open cells and removing the bulk of their cell walls and carbohydrates.

STEP ONE

1. Obtain 50 grams of diced banana (about one small banana cut into cubes not longer than 3 mm long) and two 500-mL beakers from the equipment bench.

2. Add 100 mL of homogenating solution containing sodium dodecyl sulfate (a detergent) from the equipment bench to your 500-mL beaker. Incubate the diced banana/ homogenating solution mixture in a 60°C water bath for 15 minutes to denature the banana's proteins and to dissolve its cell membranes and walls. A citrate buffer in the homogenating solution will also inhibit the action of enzymes that might damage the DNA.

3. After you have incubated your solution for fifteen (15) minutes, cool it by placing your beaker in a water bath containing ice for three (3) minutes. Transfer the cooled preparation to a blender and fasten the blender's lid. Homogenize for 45 seconds at low speed followed by 30 seconds at high speed to further break open the cells and release their contents (carbohydrates, fats, proteins, and nucleic acids).

4. Transfer the homogenated banana to your 500-mL beaker and thoroughly cool it by placing it in the ice/water bath for fifteen minutes. Transfer the homogenate into a 500-mL beaker by filtering it through four thicknesses of cheesecloth. Return the two 500-mL beakers to the equipment bench.

Once the cells have been broken open and the bulk of the cell walls and carbohydrates have been removed, the next step in a DNA purification typically involves the removal of cellular proteins and lipids.

STEP TWO

1. Obtain two clean 250-mL flasks. Pour 50 mL (and not more) of your filtered homogenate into one of the 250-mL flasks. Ask your teaching assistant for a demonstration in the proper handling of chloroform. In the prep lab's fume hood, use the technique your teaching assistant demonstrated to add 10 mL of chloroform to the homogenate by pouring it down the side of the flask.

2. Gently swirl the contents of the 250-mL flask for between 30 and 60 seconds. A cloudy white interface of denatured protein and lipids should appear between the lower chloroform and upper aqueous layers. Overly vigorous swirling may result in the formation of an emulsion in which the denser chloroform layer is no longer discernible from the lighter aqueous layer containing your homogenate—if an emulsion is formed, asked your teaching assistant for help in resolving it.

3. Carefully pour the upper, aqueous layer into your second clean 250-mL flask while leaving all of the chloroform and interface material behind. Discard the contents of the flask that still contains the chloroform and interface in the special container for this waste and clean the flask with water for further use.

4. Repeat the previous two points (2 and 3) of this step four more times. With each repetition, less denatured protein and lipid should appear at the interface between the chloroform and aqueous layers.

5. After the last chloroform extraction, transfer the homogenate into a clean 250-mL beaker from the equipment bench very gently so that absolutely no chloroform is transferred with it. Leaving behind a small amount of homogenate is far preferable to carrying chloroform over to the next step in the experiment.

6. Place the flask containing the homogenate in an ice bath to thoroughly cool its contents before proceeding to the next step in the experiment. Return the flask containing chloroform to the equipment bench and obtain a glass rod.

   Each chloroform extraction removed proteins and lipids from the homogenate and increased the relative purity of the nucleic acids within it. Once the cellular carbohydrates, proteins, and lipids have separated from the cell's nucleic acids, all that remains in a standard DNA purification is a step to concentrate the DNA through a precipitation with an alcohol.

STEP THREE

1. Once your homogenate is quite cold (less than 15°C), transfer 80 mL of very cold 95% ethanol into your clean 250-mL flask. Slowly add the cold 95% ethanol to your cold homogenate by pouring it down the side of the flask containing the homogenate until a white, stringy DNA precipitate appears (you may not need the entire 80 mL of ethanol).

2. Place your glass rod into the flask containing precipitated DNA and slowly rotate it in just one direction to "wind-up" the stringy banana DNA. Continue to rotate the rod in larger and larger circles through the beaker to collect all of the precipitated DNA.

3. If you want to keep the DNA you have isolated, fill one of the small vials on the equipment bench with 95% ethanol and transfer your DNA into it by gently easing the DNA off of the end of the glass rod. Cover the vial and seal it with a piece of Parafilm®. Barring leakage of the ethanol and extremely rough handling, your sample of purified banana DNA should remain relatively unchanged for many years.

4. Return the flasks and glass rod you have been using to the equipment bench. Please fill out the short survey at the back of your laboratory manual before leaving and give it to your teaching assistant—your laboratory work for BIO 1120 is done!

IMPORTANT QUESTIONS TO CONSIDER

Careful consideration of the following questions related to this experiment will not only insure that you understand the most important concepts of this laboratory but will also help you prepare for the lecture portion of the course.

- THE "DNA" YOU isolated in this experiment is actually a mixture of nucleic acids including a variety of RNA molecules as well as all of a banana's chromosomes. What additional steps could you perform at this point in the purification to remove all RNA from the preparation? How would you remove DNA but leave RNA behind?

HYDRONIUM ION CONCENTRATIONS AND BUFFERS, SAMPLE REPORT

INTRODUCTION

Most of our understanding of solutions is based upon work in the late 1800s by Rault, van't Huff, Arrhenius, and Gibbs. These chemists established that solutions were homogeneous mixtures of solutes in solvents. Among the most biologically important solutes found in aqueous solutions are hydronium ions (H^+). pH, the negative log of the molar concentration of hydronium ions, is the most commonly used means of describing the concentration of this solute and is the basis of a scale that ranges from 0 to 14. Solutions with pH values less than 7 are said to be "acidic," solutions above pH 7 are "basic" (or alkaline), and those that have a pH of 7 are referred to as "neutral." Even small variations in pH can have dramatic effects upon cells, and living systems typically use buffers to prevent dramatic changes in their pH.

The set of experiments described in this laboratory report use two different indicators (an anthocyanin isolated from red cabbage and others embedded in alkacid paper) to determine the pH of a variety of common household solutions through a series of comparisons to standards with pHs that range from 2 to 14. An additional indicator, phenolpthalein, was also used to demonstrate that buffered solutions are much more resistant to changes in pH than those that are not buffered.

MATERIALS AND METHODS

A pH scale using a red cabbage anthocyanin as an indicator was established as described on page 15 of the BIO 1120 laboratory manual. This pH scale was then used to determine the pH of six household solutions (Alka-Seltzer®, aspirin, beer, bleach, grape juice, and vinegar) as described on pages 19 and 20 of the BIO 1120 laboratory manual. Alkacid paper was then used to derive an independent measure of the pH of these solutions as described on page 21 of the BIO 1120 laboratory manual. The different resistances of buffered and un-buffered solutions to changes in pH was determined as described on pages 22 through 24 of the BIO 1120 laboratory manual.

RESULTS

All results are described in the attached Tables (1.2, 1.3, 1.4, and 1.5) from the BIO 1120 laboratory manual.

CONCLUSIONS

Both a red cabbage anthocyanin indicator and the indicators embedded in alkacid paper indicate that common household solutions such as bleach and vinegar have widely differing concentrations of hydronium ions (ranging from pH 4 to 13). Since living organisms are so sensitive to the pH of the solutions to which they are exposed, it is reasonable to assume that they would behave very differently in each of the six solutions that were studied. For instance, human blood with a hydronium ion concentration of 3.2×10^{-8} M is perfectly normal, but individuals with hydronium ion concentrations in their blood of 4.0×10^{-8} M suffer from acidosis and are likely to die without immediate medical attention.

The anthocyanin from the red cabbage extract showed the most dramatic and recognizable changes in color over the pH range of 4 to 14 and would make a reliable indicator for any solutions whose pH fell into that range and did not have colors that would mask those of the anthocyanin. Since the color changes of the anthocyanin at pHs below 4 involved very dark colors and tended to change with time, it is unlikely that they could be used to precisely determine the pH of very acidic solutions. In contrast, the color changes of the alkacid test strips were most readily distinguished at both ends of the pH scale and differences in the colors seen at pHs near 7 were subtle enough to introduce potentially significant errors. For those reasons, the alkacid paper would probably be the best choice for determining the pH of very basic or acidic solutions while the red cabbage anthocyanin would be more useful for solutions closer to neutral pH.

Phenolpthalein differed from the other indicators used in this experiment in that it underwent only one dramatic color change as *p*H changed from about 9 to 10. That change in *p*H occurred *much* more rapidly when sodium hydroxide was added to a solution that was not buffered than it did when it was added to a solution that had a phosphate buffer within it. Since buffers are so effective in causing solutions to be resistant to changes in their *p*H, it is likely that living systems utilize them extensively since they are so sensitive to even small changes in the molar concentration of hydronium ions.

Table 1.1 The color of a red cabbage anthocyanin in solutions of known *p*H.

*p*H of Standard	Color of Anthocyanin	Observations
2	Deep Red	Very Intense Color
4	Red	Color Became Lighter with Time
6	Pink	Seen Only against White Paper
7	Purple	Less Intense than at pH 2
8	Blue	Small Bubbles Also Present
10	Blue/Green	Color Became Darker with Time
12	Lime Green	Relatively Faint Color
14	Golden/Yellow	Large Bubbles Also Present

Table 1.2 The pH of six household solutions using an anthocyanin indicator.

Solution	Name	Color of Anthocyanin	Estimated pH
A	Alka-Seltzer®	Blue	8
B	Aspirin	Pink	6
C	Beer	Blue/Green	10
D	Bleach	Golden	13
E	Grape Juice	Purple	7
F	Vinegar	Red	4

Table 1.3 The pH of six household solutions using alkacid test paper.

Solution	Name	Color of Anthocyanin	Estimated pH
A	Alka-Seltzer®	Light Green	9
B	Aspirin	Dark Green	8
C	Beer	Blue	10
D	Bleach	Black	13
E	Grape Juice	Green	6
F	Vinegar	Brown	4

Table 1.4 Titration of an un-buffered and a phosphate-buffered solution.

Number of Drops Added	Color of Solution	pH of Solution
0	Clear	6.9
3	Pink	10.2
0	Clear	6.9
3	Clear	6.9
6	Clear	7.0
9	Clear	7.1
12	Clear	7.3
15	Clear	7.4
18	Clear	7.7
21	Clear	8.1
24	Pink	10.5

END-OF-TERM EVALUATION

EVALUATION

The laboratory manual you have used in BIO 1120 is, like our study of biology in general, a work in progress. Your opinions, comments, and criticisms of it and the laboratory in general will all be taken very seriously and used to make improvements for subsequent offerings of this course. Please take a few minutes at the end of your last laboratory section to answer the following questions about BIO 1120's laboratory and its manual.

1. Which of the thirteen experiments you performed were the most:

   a. Memorable

   b. Educational

   c. Easy

   d. Difficult

   e. Fun

2. What experiment were you most interested in telling your friends about? Why?

3. Which portion of the laboratory manual were the most:

   a. Helpful

   b. Educational

   c. Confusing

   d. Interesting

4. What changes do you wish had been made to the laboratory manual before you had to work your way through it this semester?

5. How well do you feel the laboratory portion of this course coincided with the lecture portion?

6. Do you feel that the laboratory manual was a good value?

7. How has your attitude about biology and your interest in this field of study changed as a result of your exposure to this introductory laboratory and its manual?